D0548265

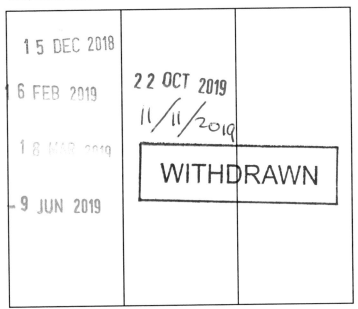
Glasgow Life and it's service brands, including Glasgow Libraries, (found at www.glasgowlife.org.uk) are operating names for Culture and Sport Glasgow

MILLION DOLLAR MATHS

MILLION DOLLAR MATHS

THE SECRET MATHS
OF BECOMING RICH
(OR POOR)

HUGH BARKER

Atlantic Books
London

First published in hardback in Great Britain in 2018 by Atlantic
Books, an imprint of Atlantic Books Ltd.

1 2 3 4 5 6 7 8 9

A CIP catalogue record for this book is available
from the British Library.

Internal illustrations © Diane Barker

Hardback ISBN: 978-1-78649-322-4
EBook ISBN: 978-1-78649-323-1

Atlantic Books
An Imprint of Atlantic Books Ltd
Ormond House
26–27 Boswell Street
London
WC1N 3JZ

www.atlantic-books.co.uk

Printed and bound in Great Britain by
TJ International Ltd, Padstow, Cornwall

Contents

INTRODUCTION

Maths and Money, A Curious Relationship

Annual income twenty pounds, annual expenditure
nineteen [pounds] nineteen [shillings] and six [pence],
result happiness. Annual income twenty pounds, annual
expenditure twenty pounds ought and six, result misery.
Charles Dickens, *David Copperfield*

Like it or not, we live in a material world, in which money
can help to create opportunities in life. We all know that
money can't buy you love or happiness. But the lack of
money can certainly lead to deprivation and frustration. So
it is only natural that people with reasonable maths skills
may occasionally ponder how that know-how can be used
to maximize their wealth. Can they, for instance, manage
their financial affairs or business better? Can they come up
with a brilliant new mathematical idea or a related piece
of technology? Or can they use their maths skills for more

nefarious purposes, such as gambling systems or hacking and cracking?

The quote above, from *David Copperfield* by Charles Dickens, points to the fact that solvency is always preferable to bankruptcy. This is not the most astounding insight, although it is sound advice in its own way. But most of us would prefer to put away a bit more than sixpence a year for a rainy day. If we're really honest, most of us would like to become as rich as possible. The self-help industry is hugely profitable largely because it sells the dream of rapid wealth for minimal effort. I won't be making that claim here, just exploring the many large and small ways that you can put maths to work for you.

I'll explore the many connections between maths and finance and the opportunities this creates for significant moneymaking. I'll include stories of famous investors, business people and gamblers who have used mathematical formulae or techniques in their work (I'll mostly avoid getting bogged down in value judgements about the morality of gambling and speculation as opposed to investment, though I will acknowledge where there are potential legal issues or other risks in a financial strategy). Modern technology also relies increasingly on maths, whether it be in the algorithms used by social media companies, the complex maths that underpins Bitcoin, or the ongoing battle between hackers, crackers and internet security experts. I'll also give quick summaries of things to do and things not to do as we go along.

The largest part of the book will focus on personal finance, gambling and investment, all of which can be easily understood by anyone with high school level maths. Some of this maths may already be obvious to you, but it is surprising how many people enjoy an occasional flutter without fully understanding the mathematics of the roulette table, or consult analytical tools such as the price-to-earnings ratio without realizing the intuitive and obvious way that this relates to interest rates. And when it comes to haggling over your salary, you may or may not know how game theory affects your chances of an increase.

Along the way we'll be meeting a strange assortment of problems that are also interesting from a purely mathematical point of view, from the Keynesian beauty contest to the Byzantine Generals Problem, and from the Kelly criterion to Maverick solitaire.

You don't need to be any kind of maths genius to put mathematical thinking to work in your everyday life. In fact most successful investors and businesspeople do not use complex maths, but instead rely on a clear understanding of how the numbers work, and of the mistakes we tend to make when we analyse data and probability. Avoiding irrational blunders can often be as crucial as making good judgement calls, and being familiar with the common mathematical and statistical errors people make is an enormous help.

It's not all easy maths – later in the book I'll discuss the mathematics of the broader financial system and maths prizes and awards, and this can't be done without an attempt to outline the more complex maths involved. It would take a far more advanced mathematician than me to have a detailed knowledge of every single maths theorem I mention. I'll be honest and own up where I'm getting out of my depth and will make it clear where the theory is likely to go beyond the reach of an amateur mathematician. But for the most part the maths required in this book is no more complex than you would learn at school.

CHAPTER 1

The Power of Exponential Growth

If a man is proud of his wealth, he should not be
praised until it is known how he employs it.
Socrates

If you ask fifty people what money is, you'll get fifty different answers: it's a peculiarly hard thing to define, so let's start by trying to pin that down. That definition will underpin the most basic ways you can make your money grow and help to explain why exponential growth is the key to successful wealth accumulation.

What is Money?

At its most basic, money is just a mathematical tool for counting and measuring value. In pre-monetary societies goods could be traded by barter in which, for instance, a sack of grain might have been swapped directly for pots or beans or for a day's labour in the fields.

Let's imagine a transaction where one dairy cow was swapped for three bushels of wheat. We could use a visual **equation** to express their **comparative value** (see Figure 1).

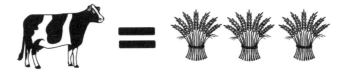

Figure 1. This represents the algebraic equation $c = 3b$ (where c represents one cow and b represents one bushel).

But you can only use pure barter if you have exactly the goods the other party wants and vice versa. Otherwise, you can end up in complicated webs of buyers and sellers where person A gives person B a cow, they give their wheat to person C, person C gives person D some beehives, and they give person A their pots and pans. This would be monstrously tricky to choreograph. So, very quickly systems of money and credit were developed. By using tally sticks or other primitive records of trades, people could sell their goods or services and store a credit to be used for purchases at a later time. If we call the unit of currency 'x', then we might have market prices of $15x$ for a cow and $5x$ for a bushel (see Figures 2 and 3).

Figure 2. One cow costs $15x$.

Figure 3. One bushel costs 5x.

In algebra, we would represent these as:

$c = 15x$
$b = 5x$

We can also manipulate these equations to get valuations for one unit of x:

$$x = \frac{c}{15}$$

$$x = \frac{b}{5}$$

Note that money can be treated as an additional item in the marketplace, whose own value can be measured in terms of other items. Its main advantage is that you can use it as an intermediary that enables transactions involving other items.

So we immediately have **counting** as the basis of monetary systems. (In fact, the whole act of counting large numbers may have been inspired by commerce – there is evidence that primitive societies would count 'one, two, three, many ...' or only up to ten or twenty, based on fingers and toes.) And we also have money being used from the start as a **measure** of comparative value.

From an early point, debt was also part of monetary systems – many societies had rules against usury (charging interest on lent money) but any system that recognizes a credit owed by one person to another already contains the concept of debt. In fact the concept of negative numbers was initially introduced by Chinese mathematicians specifically to deal with the problem of keeping accounts which recognized both credits and debits – in a ledger, the red debits were subtracted while the black credits were added.

Some people distinguish 'real money' from 'token money' or 'fiat money'. By real money they mean objects such as gold, which they see as having a real, intrinsic value, as opposed to tokens such as wooden coins or cowrie shells (which were being used as money tokens three millennia ago on the shores of the Indian Ocean). I would argue that money is always to some degree a token or representation, regardless of its physical form, but I don't want to get into the complex debate over whether gold money is more real than, say, the US dollar other than to say this: **any kind of money, whether it be gold or paper, government-backed or private, digital or a plastic token, can be valued only in relative terms**.

What this means is that the value of a unit of money can only ever be measured in terms of the goods and services (or even other currencies) it can be exchanged for.

So there is no such thing as inherent or absolute value: you can measure the current value of gold against wheat, a dollar against gold, or even the value of one yen against the value of one euro. But it is meaningless to describe any of these goods as having value in themselves without referring to who is valuing them and what they might exchange for them. All monetary values are relative and all of them fluctuate over time. And if, for instance, the price of petrol in dollars increases, it is equally valid to say that the price of dollars, as measured in petrol, has fallen.

As well as being relative, monetary value is always subjective. The same bottle of water might be worth nothing to someone who lives by a clean stream, but worth a million dollars to you if you are lost in the middle of a desert and at death's door.

The art of wealth management is based on identifying differential value and fluctuations in value. This concept is perhaps most easily understood when you consider the idea of 'net worth'. This is defined as the amount of money you would end up with if you sold all your assets and paid off all your debts **at current values**.

It can be hard to shake the idea that money does or should have an objective value. But in these days of quantitative easing (and money printing) it should be clearer than ever that money itself can gain or lose value. And it gives us a much more rigorous mathematical basis for thinking about money if we regard it simply as an item that can be exchanged for other goods and services.

What to Do

Remember that money is only a relative measure of exchange value, a way of counting the goods, services and assets it can be swapped for. At any given moment we can define the comparative value we would ascribe to two items a and b using the equation $a = nb$. And remember that the value of money fluctuates as well as the value of those goods and services. **So value is relative, subjective and fluctuating.** The main ways we can increase our wealth over time are by taking advantage of variations in value (for instance, by selling something for more than it cost us) or by adding value (for instance, creating wealth by making something more valuable out of raw materials).

Buy Low, Sell High

The next basic thing to bear in mind is that economic transactions generally rely on two individuals or groups placing a different **value** on the same item and then agreeing a mutually acceptable **price**. (If the two parties value the item exactly the same, they may agree a deal, but neither will have a strong motivation to do so.) Suppose you go out tomorrow planning to buy a second-hand car, let's say you are willing to pay up to £3,000, while the seller is willing to sell for at least £2,500. In this case a deal can usually be done somewhere in between the two prices, and this will help to set the **market price**, which is the theoretical average of many similar transactions.

The supply and demand curves (see Figure 4) that are used in economic theory are just easy ways to show how prices are set in markets. You can use mathematical tools to describe idealized versions of markets, and these are valuable analytical tools so long as you remember that the idealized markets they describe aren't actually real.

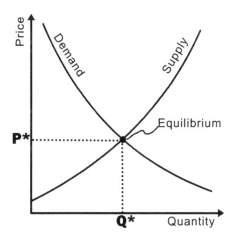

Figure 4. A supply and demand chart: As the **price** rises, **supply** tends to rise, meaning more people are willing to produce or sell an item, while **demand** tends to fall, meaning that fewer people are willing to buy the item. Theoretically the market price, or equilibrium price, will be found where the demand and supply curves meet.

Similarly if you buy a share, then this transaction works because you are assuming that the share is undervalued or valued correctly while the seller is assuming it is overvalued or valued correctly.* There may be rational or irrational reasons for these assumptions, but the key point is that the buyer and seller have different motivations and reasons for valuing these items differently and a compromise is reached. So rather than talk about 'value' it is often more useful to look at the market price, which can be measured.

If you want to make money, you have to think about the ways in which you can exchange assets, goods or services of varying price in a way that allows you to accumulate more money or possessions.

There are fundamentally four ways you might approach this task.

The first is to sell your labour for a wage or salary. In other words, get on your bicycle and go out and find some work.

The second is to create a business, large or small, in which you create goods or services. In this process you take the raw materials (whether they be labour, ingredients, materials or ideas) and create something that can be sold at a higher price. For instance, you might buy modelling clay and make brooches that you can sell for a higher

---

* Alternatively, the seller may be a 'forced seller' in which case they may believe the price to be below the fair price but have no option but to sell.

price, and advertise them online via social media to keep your costs down. By adding value to the raw materials, you are creating wealth.

The third is to invest in other people's businesses and wealth creation, either directly (by investing in the business of a friend, for instance or via stocks and shares, which you can buy directly or via a broker).

The fourth is to take advantage of the variation in value of assets, buying at low points and selling at high points – this is the basic activity of any trader who sells goods for a higher price than they pay for them, but it also describes the activities of speculators and gamblers. (It can be hard to pin down the distinction between speculation and investment, but it's worth thinking about whether the money invested is genuinely helping others to create wealth. If not, it's probably speculation rather than investment.)

However you aim to make your money, the obvious mathematical rule of 'buy low, sell high' is applicable in a world of fluctuating values. Even in the world of work, you can analyse the time and money you spend acquiring particular skills or experience and compare this to how much difference it will make to your pay. But more obviously in business and investment, the more you can take advantage of variations in value the faster your wealth will grow.

However, we shouldn't only think in terms of buying and selling. The legendary investor John C. Bogle was a great advocate of holding on to assets, writing for instance that 'the real money in investment will have to be made – as most of it has been made in the past – not out of buying and selling but of owning and holding...' In this case the

What to Do

One reason to think about value as a purely mathematical equation is that it helps us to avoid some common irrational errors. It is easy to make the mistake of considering irrelevant factors when valuing an asset: for instance, how much you paid or how hard you worked to get this asset, or how much you hoped to sell it for. This leads to errors such as the **sunk costs fallacy** (in which people find it hard to give up on a lossmaking project because of the money that has already been spent on it).

The only way you should value an asset is by considering its current value and comparing this to your other options: most of what has happened in the past is irrelevant. The trajectory of its past value may, of course, give us some information about the future trajectory, though as the adverts always say, 'past performance is no guarantee'. While the aim should always be to sell an asset for more than you paid for it, refusing to sell at a loss can be more damaging than accepting the loss and moving on.

relevant question is whether an asset is earning more than it is currently costing you to hold onto it, and how this equation compares to other assets you could swap it for. This is where the idea of comparative value is also crucial, as there is no gain to be made by selling an asset just to swap it for other less profitable assets. And the economic concept of opportunity cost refers to the fact that capital invested in one asset 'costs' us the ability to invest that same capital in an alternative asset.

The Rule of 72

When considering an investment opportunity or business model, it is often useful to know how long it will take to double your money at a particular rate of growth. (And if you're not anticipating doubling your money at some point, maybe you should be considering different, more profitable opportunities?)

The Rule of 72 is a quick way to calculate this in your head. It has been used since at least the fifteenth century when Luca Pacioli (1445–1514) included it in his *Summa de arithmetica*.

The rule is to divide 72 by the rate of growth (or the interest rate, for savings and investments): the result gives you the number of periods it will take for the initial investment to be doubled. For instance, for an interest rate of 9% a year, we divide 72 by 9 and get 8 years. The actual time it would take money to double at 9% is 8.043 years (see Figure 5), so this is reasonably accurate.

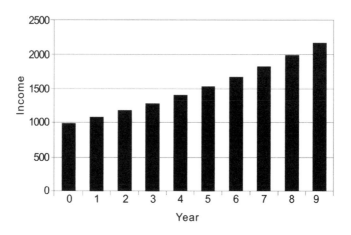

Figure 5. 9% growth per period starting with £1,000. It takes approximately eight years to double.

If you want to use this as a rule of thumb, bear in mind it is only a rough approximation, and one that works best for interest rates in the range 5–10%. Also, it's actually more accurate to use 69 or 70 as the numerator in your fraction. (72 has been used historically because it has so many factors: it can easily be divided by 1, 2, 3, 4, 6, 8, 9, 12, 18, 24, or 36.)

If you really want to get geeky about it you can use the more precise 69.3 as the numerator and use what is known as the Eckhart–McHale second order rule, which is this equation:

$$t = \frac{69.3}{r} \times \frac{200}{(200 - r)}$$

where *t* is the number of periods taken to double your money and *r* is the rate of growth. The second part of this equation helps to improve the accuracy of the estimate for high rates of growth, for which it is otherwise increasingly inaccurate.

But for most standard situations, the basic Rule of 72 is perfectly adequate, as is shown by the fact that it has served so many financiers and investors well over the centuries.

The Easy Way to Make a Million

Now that we know how to quickly work out how long it will take to double your money, let's look at an extremely simple recipe for turning an initial investment of £1,000 into a million within a year.

Let's imagine you stumble across a way to buy a supply of magic beans on Mondays. On Friday afternoon you can always sell the beans you bought for twice the price you paid for them. So you spend your start-up money on a supply of magic beans, sell them and double your money, then use the money to stock up on twice as many magic beans the next week. Hey presto, you can then keep on doubling your money – after one week you will have £2,000, after two weeks £4,000 and so on until after 10 weeks you will have £1,024,000.

I'm sure you can spot the flaw in this plan: there is no such thing as magic beans (or, more to the point, any other foolproof way of doubling your money ad infinitum). The mathematics is, however, perfectly sound. As you keep doubling n times, you multiply your original investment by the nth power of 2, so you have 2, 4, 8, 16, 32, 64, 128, 256, 512 then 1,024 (= 2^{10}) times your original investment.

So this is just basic maths, but not terribly useful in practice. However, think of it as a thought experiment in how money can grow, given a good, reliable business model. The doubling period may be somewhat longer than a week, you will undoubtedly have to work hard to find your own version of 'magic beans', and you are going to have to manage uncertainty rather than being gifted a guaranteed profit. But in the end all business plans and investment strategies rely on finding a way of making your money grow and then repeating the process.

The other thing to bear in mind is that, even if you could find a guaranteed method of doubling small amounts of money, it will become increasingly difficult to scale this up to larger amounts. For instance, if you had a system that allowed you to double your money in a casino, then it would only take a few doublings for the casino either to ban you or go out of business. Even with the magic beans, you would soon have difficulties carrying enough of them away in your wheelbarrow on a Monday. All businesses and investment systems have ceilings, but some are lower than others.

So what we need to do in this book is to focus on the grittier detail of how you might use mathematical skills and rules of thumb in combination with real-world applications to go about taking that first £1,000 and making it into £2,000. And we also need to bear in mind how much a particular strategy can be scaled up before it reaches a natural ceiling.

What to Do

When looking for your own 'magic beans', think from the start about how long it will take to double your money. And also think about how quickly this approach will hit a ceiling beyond which it is no longer possible to keep growing at the same rate.

Magic Beans in the Real World

I said there was no such thing as magic beans, and sadly there isn't. However, it is instructive to compare the magic beans business to the markets in property, land and stocks or shares. Land prices and stock markets can suffer from wild fluctuations in the short term, but in the long term they have grown pretty reliably in real terms – over decades or even centuries. So the investor or land owner who succeeds in buying at the lows and either selling at the highs or securing an income on their asset when prices rise will, in the long term, always make a good profit (as long as the long-term trend in the market continues).

What is the difference between this and the magic beans business? Firstly, there is always a degree of uncertainty about where you are in the market cycle. And secondly the cycle is much slower than the weekly interval over which I imagined magic beans doubling in value. However, there is an underlying similarity: land prices in most economies and the major markets for stocks or shares have tended to increase at about 5–10% over inflation for decades. For instance, investing in index funds (which track the performance of the entire market) will generally give this kind of return, or slightly more if you are able to enter the market in a dip. Not exactly magic beans, but a pretty good substitute for those with sufficient funds. To see how big a difference fairly small variations in annual return can make, look at these changes from 1984 to 2015 in the UK market: £100,000 invested in property over that period would have returned £502,500 (at 5.7% per annum (pa) against an inflation rate of 3.5% retail price index (RPI)), while equities (which rose at the slightly higher rate of 5.9% pa) would have returned £533,000. And constantly reinvesting the dividends from the equities could have turned that into a whopping £1,533,500 (which is equivalent to 9.9% pa). Property and land have grown more than equities since 2000, but this is largely because the property market was at such a low at that stage.

This is one of the reasons why the wealthy tend to stay wealthy (see p. 264 [the Pareto Principle]) – because these kinds of investments are most accessible for those with enough wealth to tie some of it up in long-term assets.

For those with less disposable wealth, long-term investments in property and index funds can still play a significant part in wealth accumulation but it is likely that faster methods will also be required if results are desired over years rather than decades.

The Power of Exponential Growth

The magic beans example demonstrates **exponential growth**, which means growth that continues at a constant percentage rate. This is an extremely powerful concept when it comes to wealth acquisition, and helps to explain why the wealthiest people tend to have acquired their money through investment or through owning businesses which could be successfully scaled up over time. Figure 6 shows an exponential curve in which money increases at a constant rate against time.

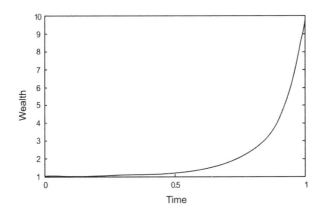

Figure 6. Exponential growth.

By comparison someone on a fairly good salary, which increases over time (but not exponentially), might see a growth in their wealth that looks more like the curve shown in Figure 7 (the vertical lines indicate pay rises).

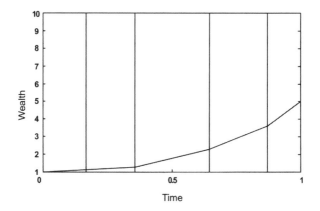

Figure 7. A slow increase in wealth as a result of pay rises.

This is a crude comparison of course, but it should be obvious that the exponential curve is the one that has the greatest potential in the long run – if you work hard and succeed, your salary might increase by a factor of 2, 10 or even 20 times over the length of your career, but to increase your earnings by 100 times or more you need to be looking for exponential growth.

So when you look at ways of making money, the very first questions you should be asking are: firstly, how long will it take me to double my money? And secondly, can this approach be scaled up, so that the money keeps growing at an exponential rate (at least in the medium term)?

Chapter 1 Summary

1. Money can be treated as a variable in a comparative value equation.
2. Use the Rule of 72 for a rough estimate of how rapidly your money will grow.
3. Exponential growth should be part of your ideal business model.
4. Unless you can find some magic beans, you will need to learn how to manage risk and uncertainty, and how to make reasonable predictions of future value.

CHAPTER 2

Beating the Casino

There can be a fine line between gambling and certain types of business, especially when it comes to speculation and investment. When mathematician Ed Thorp refined the theory of cardcounting (in which a player gains an advantage by keeping track of the remaining cards in blackjack) as a gambling strategy, his book on the subject inspired a generation of quantitative and financial analysts as well as gamblers, and he went on to become a successful hedge fund manager himself (see Chapter 5). Gambling can supply demonstrations of some basic ways of analysing luck and probability, as well as giving us an understanding of the rational fallacies that afflict gamblers. These tools and fallacies are also applicable to other, less risky investment and business options, so using gambling to explore the basic uses of maths in assessing risk and opportunity will give us a solid foundation for looking at how maths can help you to make better decisions about what to do with your money in general.

Gambling Mathematicians

The sixteenth-century polymath Gerolamo Cardano was one of the first mathematicians to lay out the foundations of probability. His book *Liber de ludo aleae* (*Book on Games of Chance*) outlined the idea of analysing events by considering all possible outcomes and how many of these were favourable to the gambler. In modern terminology, he described the 'sample space' of dice games by observing that there were 36 possible ways in which two dice might fall (see Figure 8), and, for instance, that six of these were

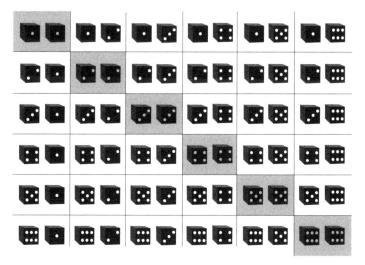

Figure 8. All possible throws with two dice.

cases where the two dice fell on the same number, of which only one was a double six. This allows us to define the probability of a double six as 1 in 36 (which in more rigorous terms means that if we throw two dice repeatedly, the number of double sixes will, over time, tend towards a limit of 1 in 36).

Dice, playing cards and gambling chips have been with us for at least a millennium and probably considerably longer. In Cardano's lifetime, the first casinos were operating in Italy. Given that running a casino relies on some knowledge of how to avoid losing money, it is possible that Cardano's achievement, which was inspired by his own daily gambling habit, was to make public the mathematical knowledge that was already being exploited for private gain, by card sharks, professional hustlers or casino operators. (A significant chunk of *Liber de ludo aleae* also focused on methods of cheating while gambling.)

While Cardano's work was a major step forwards, it was not until the correspondence between Blaise Pascal and Pierre de Fermat a century later that the theory of probability became more rigorous. Both were geniuses in their field: Pascal invented one of the first mechanical calculators, the 'Pascaline'; Fermat anticipated calculus, and Fermat's Last Theorem, which has fascinated mathematicians down the centuries, was only proved more than three centuries after his death (as we'll see in Chapter 8).

One problem that Pascal and Fermat discussed was the 'problem of points', which had previously defeated some of the world's major mathematicians. This arises when a game is abandoned before it reaches its conclusion – for instance, if a game of quoits is won by the first player to 7 points, then how should their bet be settled if the game is abandoned with the score at 6–4?

While earlier solutions had split the bet in proportion to the number of points won, or by comparing the number of points won to the winning total, Pascal and Fermat introduced the concept of **expected value** to consider all the possible outcomes – this means that we consider all the possible outcomes if this game were to be played to completion. The player with four points would need to win each of three consecutive points to win the game. We deliberately ignore the skill factor and assign a probability of $\frac{1}{2}$ to a particular player winning each individual point.

On the first point, player 1 (who leads 6–4) has a $\frac{1}{2}$ chance of winning the point and a $\frac{1}{8}$ chance of losing the point. The same applies for the second point meaning there is a $\frac{1}{4}$ chance of each of the combinations win–win, win–lose, lose–win and lose–lose. Similarly for the third point there is a $\frac{1}{8}$ chance for each of the possible combinations ranging through from win–win–win to lose–lose–lose. Since lose–lose–lose is the only way player 1 can lose, the pot should be split in a ratio of 7 to 1, favouring the player in the lead.

Alternatively we can calculate this by looking at the probability for the leading player (player 1) to win the game on any given point. This is $\frac{1}{2}$ on the first point, plus $\frac{1}{4}$ for the second point, plus $\frac{1}{8}$ for the third, which we then add up to a total of $\frac{7}{8}$.

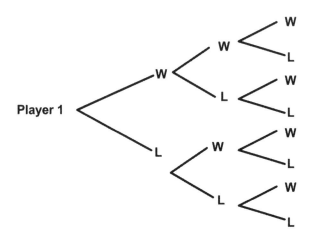

Figure 9. The sample space of the possible outcomes of the next three points played of a game – note that the diagram assumes all three points are played, but the game will actually finish earlier than that if player 1 wins the first or second point.

This process of considering all the possible outcomes in the sample space is the core of modern probability theory – Pascal's triangle is a method of calculating all the possible combinations of outcomes given particular numerical conditions. (It's a bit of a cheek that this is named after Pascal in the Western world, since it had been known to Chinese, Indian and Persian mathematicians centuries earlier.) He even took the concept of gambling to a new extreme when he introduced the idea of 'Pascal's Wager', recommending a belief in God on the basis that this is a sensible bet for a rational person to make. This rests on the assumption that the gains from not doing so (in terms of pleasure and luxury) are finite, whereas both the losses from not doing so and the gains from doing so (eternity in hell as opposed to heaven) are infinite.

Combinatorial mathematics is at the heart of modern gambling theory. For any given game or bet we can consider the sample space of all the conceivable outcomes and estimate how many of those outcomes are favourable or unfavourable. Even for gambling on the outcome of sports events, which are decided by skill rather than chance, we can estimate the odds of winning or losing as well as many other finer details, by considering historical trends and data. This is a simple process for a game of dice, but a far more complex one for a game such as poker. However, as we shall see, the art of gambling relies not just on calculation of the odds, but the search for situations in which the other party to a bet estimates the odds differently to you, so you have the opportunity to make a bet which offers genuine value.

What is a Bet?

The clearest way to conceptualize a bet is to see it as buying the right to be paid a certain amount of money in a particular future scenario. The correct way to value a bet is to analyse the sample space for all the possible alternative future scenarios. For instance, consider a bet on drawing a single ace from a full pack of 52 cards. There are four aces in the pack, so the odds on a single ace are $\frac{4}{52}$, which is one in thirteen. The question then arises as to how likely combinations of events are, and how we calculate this depends on whether the events are dependent or independent. So, if we draw a card, then return it to the pack and draw again, the two events are independent (meaning that the first draw has no impact on the probability of the second). To find the odds of drawing two aces in this way, we multiply $\frac{1}{13}$ by $\frac{1}{13}$ to get $\frac{1}{169}$. So it would be rational to bet £1 for a total return of more than £169 on this outcome. If, however, we draw two cards together, the odds change – the odds on the first card being an ace are still $\frac{4}{52}$ but the odds for the second card are $\frac{3}{51}$, so the odds for two aces drawn this way are $\frac{3 \times 4}{51 \times 52} = \frac{1}{221}$.

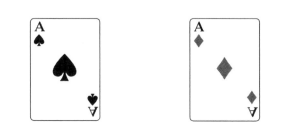

The probability of finding at least one ace after two draws from the pack (with the first card returned to the pack) are worked out slightly differently. On each draw the odds of **not** getting an ace are $\frac{48}{52}$. So the odds of getting an ace on **neither** draw are $\frac{48 \times 48}{52 \times 52}$ which we can simplify by dividing each term by 4 to $\frac{12 \times 12}{13 \times 13} = \frac{144}{169}$. To get the odds of drawing at least one ace we subtract this fraction from 1 to find the odds of $\frac{25}{169}$ (which is just over 1 in 7). And finally, as well as **multiplying** probabilities, we sometimes need to **add** them. For instance, to find the odds of drawing either an ace or a king on a single draw, we find the probability for each event ($\frac{1}{13}$) and add them together to get $\frac{2}{13}$.

In order to assess whether any given bet is good value or not, a basic grasp of probability is indispensable.

Lady Luck

Along with probability and expected value, the most important concept that a gambler needs to understand is **volatility**. Here are a few simple practical examples that demonstrate the importance of these ideas.

Firstly, imagine you have a choice of betting on one of two questions about future events. One is 'Will the town hall clock strike twelve times at noon tomorrow?' The other is 'Will this coin land on heads?'

You can bet on either. A winning bet of one dollar on the town hall clock will pay you your dollar back. A winning bet of one dollar on the coin will pay you two dollars, a losing bet will pay you nothing.

By contrast, the coin will land sometimes on heads and sometimes on tails (and both are equally probable), so this is a bet that you would sometimes win (leaving you with two coins), and sometimes lose (leaving you with zero).

In each case, to calculate the **expected value** of the bet, you do the same thing, which is to take the average win you would expect to end up with over a large number of repetitions of the same bet. Both bets have an expected value of zero, because the average return over a long period would be exactly zero for the town clock, and approximately zero for the coin toss. This indicates that both are examples of a **fair game**, meaning that neither of the parties to the bet has an advantage over the other.

However, it is clear that there is no excitement or doubt about the result of the town hall clock bet, whereas the coin toss will sometimes allow you to win some money, and other times to lose. This is because it is an event that has a volatile outcome.

So we need **volatility** for gambling, otherwise there would be no point. But the more important question that this leads on to is how to assess and calculate volatility.

For this we need to understand the statistical concept of **standard deviation**. This is a measure of how widely dispersed the members of a set are, on average. The boxed text explains how this is calculated in general.

How to Calculate Standard Deviation

Imagine ten miniature giraffes of varying heights. Here is a list of their heights in centimetres.

160, 153, 172, 159, 157, 172, 181, 177, 158, 171

First, we calculate the **mean** (average height), by adding up all these figures and dividing by the number of giraffes:

160 + 153 + 172 + 159 + 157 + 172 + 181 + 177 + 158 + 171 = 1,660

$$\frac{1,660}{10} = 166$$

So the mean height is 166cm. Next we calculate the difference of each individual height from the mean:

−6, −13, 6, −7, −9, 6, 15, 11, −8, 5

Then we square these (so that we are taking an average based on positive numbers, rather than allowing the negatives and positives to cancel each other out):

36, 169, 36, 49, 81, 36, 225, 121, 64, 25

Finally we find the mean of these figures by adding them and dividing by 10, which gives 84.2: this is the **variance** of the data set. The **standard deviation** is the square root of the variance, which is approximately 9.2cm, the average difference in height from the mean.

(The methodology becomes slightly more complex when we are measuring a sample group from a larger population, but the basic method is similar.)

For a large set of data with normal distribution,[*] we can use the **68/95/99.7 rule**. This suggests that 68% of the set will generally fall within one standard deviation of the mean, 95% will fall within two standard deviations, and 99.7% will fall within three standard deviations.[†]

---

[*] Normal distribution is by far the most common type of dispersion pattern in real life, but see also p. 42 for a look at skewed and Poisson distributions.

[†] In many cases in business and science, the three sigma rule is used as a rule of thumb, which implies that 'nearly all' members of a normally distributed set will fall within three standard deviations of the mean (and even for non-normal distributions, the proportion will be at least 88.8%).

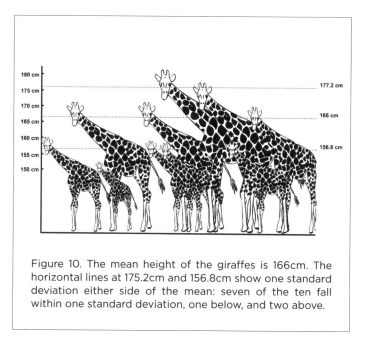

Figure 10. The mean height of the giraffes is 166cm. The horizontal lines at 175.2cm and 156.8cm show one standard deviation either side of the mean: seven of the ten fall within one standard deviation, one below, and two above.

Standard deviation effectively measures the role of luck in a given game – a large standard deviation creates a greater opportunity both for winning money and for losing it, given a rational betting strategy.

For most casino games and sports bets the standard deviation has already been calculated and published or displayed. So the most important thing is to understand how the 68/95/99.7 rule affects the outcome of betting situations in general. For a simple example, consider the following game.

Rich Man vs Poor Man

Two people set out to play a simple game of tossing a coin. They take turns calling heads or tails and the loser pays the winner £1 for each toss of the coin. This is a fair game, so over a large number of reiterations the expected value is zero for both players. However, there is an additional constraint which is that the rich man starts with £30 while the poor man starts with £10. Either is allowed go into the red for a period of time but they will be ejected from the casino if their losses exceed the starting stake after 100 throws.

As this is a binomial game (meaning there are only two possible outcomes), the standard deviation is calculated using a simple equation:

$$2 \times \sqrt{\left(\begin{array}{c}\text{number of}\\\text{coin tosses}\end{array}\right) \times \left(\begin{array}{c}\text{the chances}\\\text{of heads}\end{array}\right) \times \left(\begin{array}{c}\text{the chances}\\\text{of tails}\end{array}\right)}$$

The usual formula given for the standard deviation of the binomial distribution is the square root of [(the number of coin tosses) × (the chances of heads) × (the chances of tails)]. We have to double this because the usual assumption for the binomial distribution is based on a choice between outcomes of 0 and 1, whereas here we are looking at outcomes of 1 or −1.

So, for 100 coin tosses, the standard deviation is

$$2 \times \sqrt{\left(100 \times 0.5 \times 0.5\right)} = 10$$

Following the 68/95/99.7% rule we would expect that after 100 tosses there is a 68% chance that the profit for either player will be between −£10 and + £10.

However, this means that there is a 32% chance that the profit or loss will be larger than this. And on half of these occasions the loss would be borne by the poor man, meaning that he has about a 1 in 6 chance of being ejected from the casino at the end of the 100 tosses.

By contrast, the rich man's initial pot has a value of 3 standard deviations. We can estimate that 99.7% of the time the profit or loss will be within 3 standard deviations, so less than £30. On half of the remaining 0.3%, the loss of £30 or more would be incurred by the rich man, so we get a probability closer to 1 in 650 for the rich man to be ejected after 100 coin tosses.

This means that the rich man's odds of not being wiped out are much, much better than the poor man's, even though this is an entirely fair game. Which will come as no surprise to cynics, but it's nonetheless interesting to understand one of the mechanisms by which this comes to pass.

(Bear in mind that if the rules are different and the poor man is out of the game as soon as he loses his £10, his chances are even worse, as scenarios in which he has lost less than £10 after 100 tosses, but his losses exceeded this in the meantime, will need to be included.)

What Not to Do

Never gamble without understanding how standard deviation affects your chances of losing your pot. The formal way of measuring this is called 'Risk of Ruin' – the calculation is complex but you can find useful Risk of Ruin calculators online that will take the details of a particular gamble and output the likelihood that the starting pot will be lost in a certain period of time. Even though they are the 'rich man' in this scenario, this is as true for casinos and bookmakers as it is for bettors – they employ teams of maths geeks and statisticians to calculate how variance and volatility in games affects their chances of massive losses, in order to insure against disaster.

Rich Man, Poor Man in the Casino

Next, imagine a variation on the game above, in which the rich man and poor man are forced to play their game in a casino that takes 10% of each win. Over 100 coin tosses, the expected value for each player is now a loss of £10, so it is harder for either to make a profit.

It's obvious that there is a greater than 50% chance that the poor man will have lost his starting pot of £10 after 100 tosses – because this will happen on any occasion when he doesn't win more than 50 of those coin tosses.

However, the rich man is also affected. We know that 95% of the time the result will be within two standard deviations, meaning a profit or loss of less than £20. So about 5% of the time the profit or loss will exceed this, and on half of those the rich man will be ejected after 100 tosses, which means he now has worse than a 1 in 40 chance of ruin at that stage.

Furthermore, we can see that after 400 coin tosses, the two players' combined expected value is a loss of £40, which is the total of both original pots added together. And this is the kind of calculation that will affect your chances of winning whether you are playing a simple game like this or gambling against the house in a casino.

Which, in short, is why the business of running a casino is a very profitable one, and gambling in them is a mug's game.

Volatility and Gambling Strategies in Roulette

Most games allow you to use strategies which can increase or decrease the volatility of your betting. When you are playing roulette you will expect a higher percentage of wins betting on red or black ($\frac{18}{37}$ on a single-zero wheel) than on single numbers ($\frac{1}{37}$). But you can also use combinations to reduce the volatility further. For instance, if instead of betting £1 per round on red, you choose to bet £0.50 on red and £0.50 on odd numbers, there are 10 numbers which will win £1, 11 that will lose £1 and 16 that will break even. If you want to go even further, you could bet on reds and evens, in which case there are just 8 numbers that will win you £1, and just 9 that will lose £1 and 20 that would break even. Of course by reducing your volatility in this way, you also reduce your chances of beating the house edge. If you are the sort of gambler who accepts that you are going to lose but still enjoys the game, this kind of strategy can make your game last longer. By contrast, if your main motivation is the possibility of a big win, you could take the opposite approach of betting on the most risky outcomes and exposing yourself to the house edge as few times as possible.

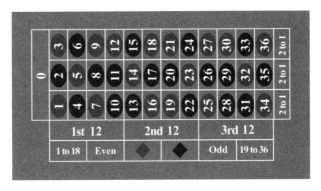

Figure 11. The standard layout of a roulette table.

Beyond Normal Distribution

When looking at any set of data (for instance the sample space of results in a casino game or a collection of sports statistics) it is important to understand what sort of distribution pattern it has. The most common pattern is called 'normal distribution' (or 'Gaussian distribution' after the brilliant mathematician Carl Friedrich Gauss). This is where the data is grouped around a central point with no particular bias to the left or right, and the mean, median and mode are all in more or less the same position (see Figure 12). This creates the familiar 'bell curve' in which most individual elements of the set are close to the average, with outliers becoming increasingly rare above and below this point. The 68/95/99.7 rule (see p. 34) works best for this kind of distribution.

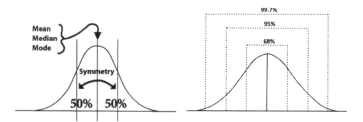

Figure 12. A bell curve showing normal distribution. Given normal distribution, we should expect 68% of the data set to fall within one standard deviation (indicated by the vertical lines either side of the mean), 95% within two and 99.7% within three.

However, it is worth bearing in mind that not all sets of data have such a simple distribution, in which case any analysis of expected value based on calculating the mean value of the set becomes less useful. In particular, it's worth being aware that some sets of data produce a graph which is skewed to the right or left, as in Figure 13. In this case, the mean may be at a different point of the curve to the median.

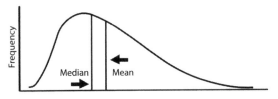

Right skewed distribution: Mean is to the right

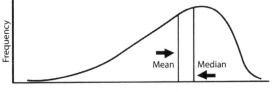

Left skewed distribution: Mean is to the left

Figure 13. A right skewed distribution has the mean to the right of the median. A left skewed distribution has the mean to the left of the median.

Consider a lottery as an example of a highly skewed distribution. Let's assume there is one jackpot of £100,000 plus 50 prizes of £2,000 each, and 250,000 tickets have been sold at £1 each. The mean return achieved by buying a ticket is actually £0.80, but the median and mode are both zero and this is by far the most likely return given that the chance of winning anything at all is only 51 in 250,000. Of course, people still buy lottery tickets in the hope of winning the big prize, regardless of how the odds are stacked against them – indeed the high volatility involved in a game like this is a large part of the appeal – people know that losing is the most common result, but the excitement of imagining a huge win overcomes that rational doubt. For the same reason, many people who play slot machines prefer the ones with high volatility – large but less frequent payouts – because they find it easier to believe they will come out ahead on these machines.

Another type of skewed distribution that is worth knowing about is the Poisson distribution. The easiest way to think about this is that it generally applies when we are counting events that are spread out over time periods and for which the smallest number we can count is zero. If we are told, for instance, that there is an average of six emergency admissions each day at a particular hospital, a Poisson distribution will give us a good estimate of how likely it is that on any given day there are zero admissions, one admission, two admissions, three admissions and so on. This is generally displayed as a bar chart, or a table of the relevant values.

To explain the difference between a normal distribution and a Poisson distribution, I've collated a set of statistics for my local soccer team Arsenal's performances in 2016–17. First, let's look at goal difference.

Figure 14 is a bar chart of the differences in the number of goals scored in a particular game, with the height of the bars proportional to the number of matches that ended with a specific result from –4 (two dreadful 5–1 thrashings by Bayern Munich in the Champions League) through to + 8 (an 8–0 win over Viking FK in a meaningless friendly match).

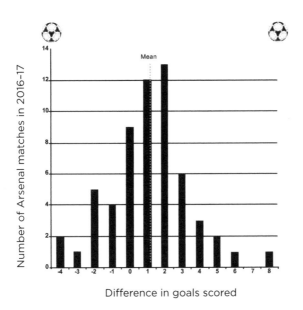

Difference in goals scored

Figure 14. Difference in goals scored for Arsenal matches 2016-17.

The average (mean) goal difference was 1.13, and you can see that the chart follows roughly in the shape of a bell curve, with the mean falling between the two tallest columns, 1 and 2. There is no particular reason to expect goal difference to have a skewed distribution as a wide variety of values are feasible and it is not bounded at zero. One would therefore expect goal difference for all teams across a season to tend towards the bell curve.

When it comes to the number of goals scored by Arsenal, we get a different looking graph (see Figure 15), mainly because this statistic is bounded at zero (meaning it can't contain any values less than zero) and the median number of goals will often be lower than the mean, since a few high-scoring matches will push the mean higher. Notice that the curve stretches out further to the right than to the left, because of those few high scores.

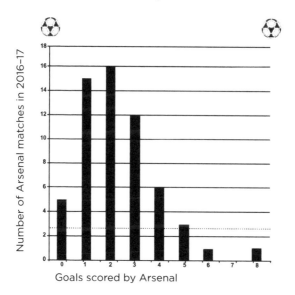

Figure 15. Number of goals scored by Arsenal in 2016–17 matches.

The average number of goals was 2.3, which for a normal distribution might lead us to expect more games with three Arsenal goals than one, since the mean is closer to 3 than 1. It is in situations like this that the Poisson distribution is our friend. Using an online Poisson calculator we can input the average of 2.3, with an upper limit of 8 goals, and it generates the probability values shown in Table 1. Applying these to the total of 59 games we get the estimate in the third column – which is not perfect, but pretty accurate, and correct in so far as it estimates that there were more games where Arsenal scored one goal than three. Paying too much attention to the mean number of goals scored in this case could therefore give you a misleading idea of the actual probabilities involved.

Goals	Probability	Rounded estimate of number of games
0	0.10	6
1	0.23	14
2	0.27	16
3	0.20	12
4	0.12	7
5	0.05	3
6	0.02	1
7	0.01	0
8	0.001	0

Table 1. Poisson distribution of number of goals scored by Arsenal 2016–17

One way to use this kind of statistic is to take the mean of goals scored for each of two teams across a recent period, generate a Poisson distribution of goals for each team in order to extrapolate how frequently they are currently likely to score, and to use that to test the probability offered by the bookmakers against your own estimates.

If for instance we concluded that Arsenal's most common score of two goals has a probability of 0.27 as in Table 1, and their opponents Spurs' most common score of two goals has 0.25 probability, we could estimate the likelihood of a 2–2 draw as $0.27 \times 0.25 = 6.75\%$. If the bookmaker happens to be offering 20 to 1 (digital odds of 21) on this outcome, we might be tempted as the price would be good value when compared to our estimate. (See section 'What are the Odds?' below.)

We could further hone this by narrowing down our initial search to home and away games, to matches in a particular competition, to games against the better teams and worse teams and so on. Obviously the quality of the input data will affect the output, and this is only one way of estimating probability, but these are the kinds of tools that we can use to get a clearer understanding of the real patterns behind stats.

It's worth noting that when we apply a Poisson distribution to sets of data with a higher mean it can appear more similar to the normal distribution – the closer the mean is to zero, the more the graph will skew to the right. Figure 16 illustrates this, by reiterating Poisson curves for an increasing mean. But any measure of events in time bounded at zero is nonetheless likely to be better modelled by Poisson distribution than normal distribution.

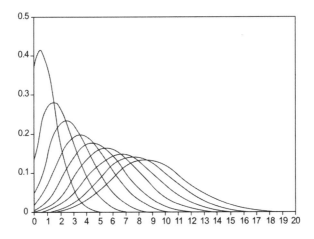

Figure 16. Poisson curves for a variety of means.

What are the Odds?

When considering the odds offered by bookmakers and casinos, it is often important to be able to calculate the probability that they imply. You also need to be able to convert between the different formats in which odds can be offered. The most common odds in Britain are fractional odds, while in Europe, Australia, Canada, New Zealand and other parts of the world digital (or decimal) odds are more common, and in the USA Moneyline (or American) odds are often used. In this book we will use fractional odds, with the decimal odds in brackets where necessary.

Below is a breakdown of how to interpret each of these and how to convert each into an implied probability. The general rule for calculating implied probability is to divide the stake by the potential payout (and for probability expressed as a percentage, multiply by 100), but the precise method differs for the three systems.

FRACTIONAL ODDS

In this system, the odds are quoted as two numbers 'x to y', written as a fraction with x over y, for instance 7/1 (or seven to one). y represents your stake, while x represents the winnings. So a winning bet of £1 at 7/1 will return your stake of £1 plus £7 winnings, £8 in total. A winning bet of 4/6 (four to six, or 'six to four on') will return £10 in total for a stake of £6, meaning you have won £4.

To calculate the probability that fractional odds would imply in a fair game, divide 100% by $(x + y)$, then multiply by y. So odds of 7/1 theoretically imply that the book-maker or casino believes this event has a probability of 100/8 = 12.5%, while odds of 4/6 imply a probability of $(100/10) \times 6 = 60\%$.

DECIMAL ODDS

With decimal odds, your original stake is assumed to be worth 1, and the odds tell you how much you would win in total for that bet. So digital odds of 8.00 are equivalent to 7/1 in fractional odds, while digital odds of 1.66 are roughly equal to 4/6.

To convert decimal odds to an implied probability, simply divide 100% by the odds. For instance digital odds of 4.00 would imply a 25% probability in a fair game.

AMERICAN ODDS

In the Moneyline or American system, odds are quoted as a number preceded by a plus or minus symbol. A minus symbol indicates odds that are lower than evens (in the fractional system) or 2.00 (in the decimal system), indicating that this is one of the most likely (or favourite) outcomes, while a plus symbol implies an outcome that has a probability of less than 50%. For odds with a minus symbol, the number indicates how much you would need to bet to **win** 100, while for a plus symbol the number indicates how much you would win if you **stake** 100. For instance odds of –350 mean that a $35 dollar stake would win $10 (making a $45 dollar return) while odds of +225 mean that a $1 dollar stake would win $2.25, making for a $3.25 return.

To convert to percentage probability, divide 100% by the total return, then multiply by the stake (ignoring the minus symbols). So for –350, the sum should be $(\frac{100\%}{450}) \times 350$ (approximately 77.7%). And for +225, the sum should be $(\frac{100\%}{325}) \times 100$ (approximately 30.77%).

The search for a value bet is essentially about finding bets where you believe the true probability of an event is higher than the implied probability of the odds on offer. However, one of the reasons this is so unusual is that the bookmakers and casinos are not playing a fair game, but instead are building in a profitable margin to the odds they offer – so the odds on offer are already worse than their estimate of the true probability. So next we need to look at how to calculate house edge and the effect it has on gambling situations.

The House Edge

Running a casino, game or betting book wouldn't be a profitable enterprise if the odds weren't somehow loaded in favour of the house and against the gambler. Earlier we looked at examples such as betting on a coin toss with odds of evens (2.0) for a winning bet, meaning that each bet had an expected value of zero. In practice, unless you are gambling among friends with no-one running the game, your gambling opportunities are likely to have an expected value of less than zero, meaning that you should be expecting to lose.

The house edge (also known as the cut, the take, the juice, the underjuice or the vigorish) is defined as the casino's profit, expressed as a percentage of the original bet. For most games the way to calculate this is to analyse the range of probabilities across the sample space for that game. For instance, when playing roulette with a single zero, there are 37 possible outcomes. If you bet £1 on red, there are 18 winning slots and 19 losing slots on the roulette wheel, each of which is equally probable. So your expected value is

$$\frac{18}{37} - \frac{19}{37} = \frac{-1}{37}$$

Converted into percentage terms, this is an expected value of −2.7%,* meaning that the house has a positive expected value of 2.7; and since other betting options for roulette rely on the same basic maths, this is the house edge for this game.

Similarly for roulette with two zeros, the winning versus losing bets equation is

$$\frac{18}{38} - \frac{20}{38} = \frac{-1}{19}$$

Here the house edge is 5.26%.

If we imagine drawing cards from a pack of 52 cards, and being offered odds of 10 to 1 (11.00) if we correctly guess the value of the card (in other words whether it is an ace, two, king, queen, etc.), we would expect $\frac{4}{52}$ winning bets and an average return of £44 for every £52 gambled. Here our expected value is $\frac{-8}{52}$ and the house edge is thus 15.4%.

A more general expression for house edge is to calculate the expected value as the following addition:

$\sum$ [(probability of event i) × (value of the return of event i) for all events i]

The house edge is the negative of the expected value, the percentage of the initial bet that the house will keep, on

---

* Rounded to one decimal place – in general I will be rounding such calculations to one or two decimal places without always drawing attention to this.

average. (A complication arises where ties are feasible within the game – in this case the general rule is only to calculate the expected value across the range of bets that will be won or lost, not tied.)

We can also look at this in terms of implied probability. The house or bookmaker create their edge by calculating what they think the implied probability is for an outcome, and then pricing it as though it is slightly **more** probable than that (thus offering shorter, less generous odds). So another way to think about house edge is that it is created by the bookmaker acting as though the range of possible outcomes adds up to more than 100% probability in total.

So here's how to calculate the edge when betting on sports events. First identify all the possible outcomes in a class of bets. For instance on a five-horse race the odds might be as below:

🐎	Nimrod's Son	4/5	(1.80)
🐎	Roadrunner	4/1	(5.00)
🐎	Faded Glamour	6/1	(7.00)
🐎	Rilkean Heart	10/1	(11.00)
🐎	Electric Dreams	10/1	(11.00)

Let's make a list of how much you would need to bet on any given horse to win a return of £100 total. To do this divide £100 by the decimal odds (or by 1 more than the fractional odds).

	Nimrod's Son	$\frac{£100}{1.8}$	= £55.56
	Roadrunner	$\frac{£100}{5}$	= £20.00
	Faded Glamour	$\frac{£100}{7}$	= £14.28
	Rilkean Heart	$\frac{£100}{11}$	= £9.09
	Electric Dreams	$\frac{£100}{11}$	= £9.09

If we placed each of these bets we would be guaranteed a total return of £100. However, we would have spent a total of £108.02 to achieve this. (And since this is also how we would calculate the implied probability for each outcome, the implied probabilities for each of the five horses add up to 108.02%.) So our expected return is $\frac{100}{108.02}$ = 92.6% (expressed as a percentage). The bookmaker has an edge of 7.4%, and for every £1.00 gambled we would expect an average return of less than £0.93.

Finding the House Edge

Of course, while it is useful to understand the principles, you don't always need to do all the calculations to discover the house edge yourself. Casinos and bookmakers are mostly obliged to publish and display the house edge for particular markets, games and machines, and these are widely available in books and on websites. A house edge of 5–10% is fairly common, but slot machines, lotteries and Keno can have a house edge as high as 15% or even 25%. If you have learned how to use an optimal strategy, blackjack has a much lower initial edge, just 0.5%, which is derived from the fact that the dealer wins the bet when players go bust even if he goes bust himself. Some video poker machines have a 0.5% edge, while others are far less generous – poker played in a casino is not directly subject to an edge within the game, but the house generally takes a regular cut which achieves the same result in a different way.

So given that all these games are rigged against the gambler, why do people still play these games of chance?

Good and Bad Reasons for Gambling

First, let's acknowledge that many of the reasons for gambling are simply irrational. People aren't terribly good at understanding and assessing risk. Some gamblers believe that in spite of the odds being rigged against them, they are simply luckier or otherwise immune to the cold logic of the maths. Many gamblers remember their wins more clearly than their losses, or only count their winnings from the end of their last losing streak, and thus have inaccurate perceptions of how well they are doing. Other gamblers see the ups and downs of a gambling session and believe that they will be able to quit while they are ahead. (This is known as 'gamblers conceit' and is generally a fallacy as it is all too easy to see winnings as 'free money' with which to pursue greater success.)

None of these are good reasons to gamble, and it is worth being aware of the dangers of muddled thinking about the subject (for more on gambling fallacies, see p. 98).

Gambling addiction is also a very serious problem, which leads to lives being destroyed around the world.

However, there are also many gamblers who do understand that they lose money, but simply enjoy the process as a pastime and set aside a sensible amount of money that they can afford to lose.

Next it is worth mentioning that some people who understand the maths of gambling, and can work out house edges and standard deviations for particular games, nonetheless misconstrue the consequences, and believe they can beat the odds. For instance, you will often hear standard deviation being described as though it is the gamblers' friend, and a measure of how lucky you can be in a particular game. This is misleading, and as much of a delusion as some of the more obvious gamblers' delusions – standard deviation does measure how close to the expected value you are likely to end up in a gambling session, and it is true that a game with zero volatility offers zero opportunities for gambling. But always bear in mind that standard deviation shows you both how much better off **and how much worse off** you might be than the expected value, and that both are equally probable.

Also bear in mind that standard deviation is proportional to the square root of the bet value, while expected return is proportional to the full bet value. This means that the longer the gambling session, the less likely it is that the gambler will be able to beat the house as the rate at which you should expect to lose money is larger than the rate at which luck can help you out. This is also a reason for not being too timid in your gambling strategy – the volatility of a single bet is far higher than the volatility of a thousand bets: the more bets your session is divided into, the more likely you are to lose money at a rate close to the house edge (see the law of large numbers, p.89).

In the end there are lots of reasons to be sceptical about the idea of making money by gambling. The only convincing reason to attempt this is if you have a valid reason to believe that you can neutralize or overcome the house edge. This can happen, for instance, within a game of blackjack – the edge varies through the game as the make-up of the remaining cards changes, and there are situations in which the edge can tip in favour of the expert gambler.

The Inverse Rule of 72 (and Managing Your Bankroll)

First, let's take a look at a few ways you might lose your money while gambling. Firstly, **gambler's ruin** describes the way that a persistent gambler who increases his stake when he is winning, but fails to reduce it when he is losing will inevitably lose all his money. Blaise Pascal and Pierre de Fermat discussed this problem in the seventeenth century, and the mathematician Christiaan Huygens formulated a general way of calculating the probability of each player winning a series of bets which will conclude when one player loses his bankroll. (The concept of gambler's ruin is also used to refer to the idea that, when playing a fair game, a gambler with finite wealth will inevitably lose all his money sooner or later when playing a gambler with infinite wealth – this can be proved by modelling the situation using a random walk, in which a series of consecutive moves up and down are chosen at random, on the number line.)

However, all we need to do here is to look at the most basic version of the idea. Imagine a gambler who increases his bet in proportion to his bankroll when he is winning, but fails to reduce it when he is losing. If he starts out betting n on each bet, and doubles his bankroll, he increases his stake to $2n$.

If this is a fair game, it is equally possible that the gambler will go broke or double his money. So we can allocate a $\frac{1}{2}$ chance to him going broke before he doubles his money. If, however, he doubles it, then we can repeat the same scenario. From this point he also has a $\frac{1}{2}$ chance of doubling his money and a $\frac{1}{2}$ chance of going broke. So the chance of him going broke before he doubles his money twice is $\frac{1}{2} + \frac{1}{4}$. And at each doubling, this process continues, meaning that after m rounds the probability of him going broke is

$$\frac{1}{2} + \frac{1}{4} + \frac{1}{8} + \frac{1}{16} + ... \left(\frac{1}{2}\right)^m$$

This tends towards a limit of 1, so the chance of him going broke eventually tends to 100%, in spite of this being a fair game.

This may seem academic, but it is actually a common problem for gamblers, who recognize that scaling up is the best way to win exponential amounts, but fail to apply the same thought process in reverse. And it is actively encouraged by many casinos who sneakily 'chip up' when gamblers win, giving them larger-value chips so that they will increase their stakes.

However, now let's look at a more sensible gambler who does reduce their stake when they lose, and a more realistic scenario in which this isn't a fair game, but there is a house edge of 6%. While it is not a perfect rule of thumb, we can use an inverse version of the Rule of 72 that we learnt in the first chapter to estimate how long it will take for the gambler to lose half their bankroll. Dividing 72 by 6 gives us 12, which suggests that on average, it would take bets to about 12 times the value of the full bankroll to lose half of it. As this is repeated the bankroll will eventually reduce to an amount too small to make a valid bet, since money isn't infinitely divisible.

However, the smaller the proportion of the bankroll the gambler risks on each bet, the more bets it will take them to lose their money. So this shows us the importance of managing your bankroll as well as looking for value betting opportunities.

Chapter 2 Summary

1. If you enjoy gambling, make sure you are as honest with yourself as possible about your motivations and fallacies.

2. Given the huge advantage you are conceding to the casino or bookmaker simply by playing their games, the best way to gamble is not at all, and the second best way is in small quantities that you enjoy for recreational purposes.

3. For any given bet, you can attempt to calculate the expected value. In a fair game, your expected profit will be zero.

4. Standard deviation measures volatility. The more volatile a betting scenario is, the more chance you have of either winning or losing.

5. Where there is a house edge, as with casinos and bookmakers, your expected value is always negative and it is important to understand your risk of ruin.

CHAPTER 3

Gambling Systems
and Strategies

A gambler never makes the same mistake twice. It's
usually three or more times.
V. P. Pappy

It is common for gamblers to use a variety of systems and
strategies to try to improve their chances of a profit. These
include optimal betting, Kelly betting, hedging, value
betting and betting systems. Let's take a look at these to
try to assess how helpful they might or might not be.

Foolproof Systems

Over the years many gamblers have believed they have
invented a perfect betting system which removes all the
risk from the process and guarantees a profit. And many
hucksters and snake oil salesmen have claimed to have
invented such a system in order to fleece gullible punters.

So the first thing that needs to be said about betting systems is that they are indeed foolproof. Because being taken in by one is a foolproof way for you to ensure that the casino or bookmaker wins money from the punter in the short, medium and long term.

To understand why betting systems are generally flawed, let's first look at one of the best-known systems in history.

THE MARTINGALE SYSTEM

This notorious system involves starting with a single unit bet on an evens bet such as red or black at the roulette table. If you win, you bank the winning unit and start again. If you lose you double your bet for the next round. So a win on the second round will return 2 units winnings, 4 units in total. Since you have staked 3 units, this is a total profit of 1 unit. At this point you restart the process.

For any completed cycle you come away with 1 unit winnings, whether it be 8 units returned for $1 + 2 + 4 = 7$ units on the third round, 16 units returned for $1 + 2 + 4 + 8 = 15$ units staked, and so on. As a general rule, after n rounds, you have bet $(2^n - 1)$ units and can hope to win a return of 2^n units.

Of course the problem is that your stake is increasing exponentially, so all it takes is a run of n losses, where $[2^{n+1} - 1]$ is larger than your initial bankroll, to wipe you out. For instance 10 losses in a row would add up to $1 + 2 + 4 + 8 + 16 + 32 + 64 + 128 + 256 + 512 = 1,023$ units staked. If you started with less than 2,047 units, you can't now continue with the system. If you started with 1,024 units then you have only 1 unit left, which you will need to hang on to, so you can buy a drink to drown your sorrows.

The key thing to notice about the Martingale is that it concentrates all the losses into one relatively improbable outcome (but hugely increases the loss made when that outcome arrives). The expected return from the system is exactly the same as it is for simply betting 1 unit on every round.

To make this clear, let's look at every possible outcome in a fair game (in other words we will ignore the possibility of landing on zero, which is where the house gets its edge from) for three rounds of play, where we are betting on red.

RRR wins 3 units
RRB wins 1 unit
RBR wins 2 units
RBB loses 2 units
BBB loses 7 units
BBR wins 1 unit
BRB breaks even
BRR wins 2 units

The potential wins and losses are equal, but most of the potential losses have been concentrated into the BBB outcome. There is a $\frac{1}{8}$ probability of this outcome, which will lose us 7 units, while we have a $\frac{7}{8}$ probability of winning an average of 1 unit for the other sequences. As we increase the number of rounds to the point where the potential losses outweigh the bankroll, this pattern continues and there is no overall advantage gained. And, of course, in a casino the house edge shifts the actual probability of winning to the advantage of the house. People's faith in this system is largely based on a misunderstanding of how common long streaks of the same result are, even within a random chain.

This is fundamentally how all betting systems work – they manipulate the probability of a winning session, but only by increasing the cost of a losing session. To do this they either concentrate all the risk into a small area of the sample space, or only increase the chance of a regular biggish win by spreading the chance of a slightly increased loss out to balance this. What they can never, ever do is to eliminate the chance of a losing session, or alter the fundamentals of expected value and house edge.

THE 1–3–2–6 SYSTEM

For another example, consider the 1–3–2–6 system, which is also used for evens stake betting on roulette. This is very simple and has the advantage of not increasing your stake exponentially as in the Martingale. All you do is choose a betting unit (commonly recommended to be about $\frac{1}{50}$ of your bankroll) and start with a bet of a single unit. If you win, you increase the bet to 3 units. If you win again, decrease to 2, and if you win for a fourth time increase it to 6 units. Then you start the process over again with the same initial stake. If at any point you lose, you go back to the start of the cycle.

As with the Martingale, people's faith in the system is based on a misunderstanding of the probability. Here, we could naively reason thus. There are five possible outcomes.

First loss comes on the first bet. (Lose 1 unit.)
First loss comes on the second bet. (Lose 2 units.)
First loss on the third bet. (Win 2 units.)
First loss on the fourth bet. (Break even.)
Win all four bets. (Win 12 units.)

This might look like we have a one in five chance of winning 12 units, and a four in five chance of a small win or loss. But those aren't the true probabilities. Instead we should analyse it thus (bearing in mind we are again assuming a fair game):

First loss, first bet. (Lose 1 unit: $\frac{1}{2}$ probability)
First loss, second bet. (Lose 2 units: $\frac{1}{4}$ probability)
First loss, third bet. (Win 2 units: $\frac{1}{8}$ probability)
First loss, fourth bet. (Break even: $\frac{1}{16}$ probability)
Win all four bets. (Win 12 units: $\frac{1}{16}$ probability)

So our expected return in a fair game is $(-0.5 - 0.5 + 0.25 + 0 + 0.75) = 0$.

Thus we have magnified the win for a four-bet winning sequence but failed to increase our chances of winning overall. The system may provide some fun for people using it and make the gambler feel good by providing reasonably regular wins of 12 units, and it is not as potentially ruinous as the Martingale – but on the whole it is not rational to use it.

THE LABOUCHERE SYSTEM

Let's hope the message is clear at this point: betting systems simply don't work, no matter how intuitively strong they may appear. There are a wide variety of progressive betting systems out there that we could analyse and dismiss, including the Fibonacci system (which keeps increasing the stake, but in accordance with the Fibonacci series), the D'Alembert and the Paroli. All of these rely on the same basic flawed logic as the Martingale, which has bankrupted countless gamblers throughout history, but vary in the degree to which the stakes are increased. Rather than give chapter and verse on all of these systems, let's take a look at the Labouchere (or cancellation) system, which has the merit of being mathematically interesting and the demerit of being persistently hawked on the internet as an infallible method by people who should know better.

The basic system is that you start with a piece of paper and write down a series of gradually increasing numbers, as below:

0, 0, 1, 1, 1, 1, 2, 2

The sum of these, 8 in this case, is the amount we are trying to win in one cycle of the system. (You can use any alternative series of increasing numbers, depending on how much you want to win and how fast you want to escalate the risks.) As with the previous systems, this is mainly for evens bets such as red or black in roulette, although adaptations of it have been used for a variety of other games and for sports betting.

For your first stake, add together the first and last number on the list and bet this number of units. Here, that would be 0 + 2 = 2 units. Then, if you win, cross those two numbers out thus:

0, 1, 1, 1, 1, 2

Or, if you lose, add the total of the losing bet to the end of the list:

0, 0, 1, 1, 1, 1, 2, 2, 2

You repeat this process until you have either crossed out all the numbers (in which case you will have won the 8 units of the original list) or have one number left, in which case use that as the stake for your next bet. If that wins, you have completed a cycle. If not, you can just continue the process.

For instance here is one way this process might lead to a 15-unit win. The result of each bet is followed by the amended list and the balance. Notice that at every stage the total of the remaining list and the balance add up to 15 units.

Initial list 1, 2, 3, 4, 5 (Target Win = 15)

Bet 6, Lose

1, 2, 3, 4, 5, 6 (–6)

Bet 7, Lose

1, 2, 3, 4, 5, 6, 7 (–13)

Bet 8, Win

2, 3, 4, 5, 6 (–5)

Bet 8, Lose

2, 3, 4, 5, 6, 8 (–13)

Bet 10, Win

3, 4, 5, 6 (–3)

Bet 9, Lose

3, 4, 5, 6, 9 (–12)

Bet 12, Lose

3, 4, 5, 6, 9, 12 (–24)

Bet 15, Win

4, 5, 6, 9 (–9)

Bet 13, Win

5, 6 (+4)

Bet 11, Win

End (+15)

The first interesting thing about this sequence of bets is that we have had 5 wins and 5 losses, but the system has nonetheless produced a profit. And note that if we now start again with a fresh bet of 6 and lose, we will still be up by 9 units in spite of having made more losing bets than winning ones. (This can happen because in this case the winning bets are larger on average than the losing ones.) This kind of outcome can lead people to believe that it is a system that somehow overcomes house edge and the odds to produce a guaranteed profit.

So let's work through all the possible outcomes from the first three rounds (using the list 1, 2, 3, 4, 5 as above) to understand what is happening here. We'll use the abbreviation W6 (+12) to mean 'Win 6 units, balance after round = +12' and so on, as shown in Figure 17.

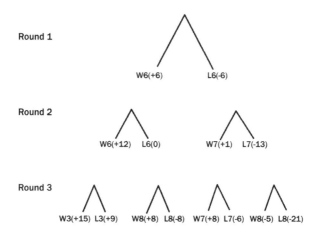

Figure 17. Outcomes from the first three rounds of the Labouchere system.

The first thing to note is that the total of all the possible outcomes at the end of each round is zero. Since each outcome is equally probable in a fair game, our expected value is zero. And this will always be true after any number of rounds, since the win and lose options for any given branch will add up to zero.

So how has the illusion of advantage been created? Well, the winning options (+15, +9, +8, +8) have been flattened out and bunched together, while the losing options (−5, −6, −8, −21) have been stretched out, with a large part of the potential losses concentrated into the worst outcome (−21). If you played this game for 10 rounds, repeating from the start after a win, you would win 51 units for 10 wins in a row, but at the cost of potentially losing 105 units for 10 losses in a row. It's not as extreme as the Martingale, but pretty much the same trick is being played. And no matter how you fiddle with the starting list of numbers, the basic maths won't change.

Some fans of this system point out that by adding one number to the list for a loss, while deleting two for a win, you would need less than one win for every two losses for the strategy to fail. But this ignores the law of large numbers (see p. 89) which makes it clear you shouldn't assume such outcomes are particularly rare, and also ignores how easily you can run out of cash if you rely on a system which gradually increases your stake during a losing streak.

The Reverse Labouchere also has its advocates. Here you reverse the operations in the hope that 'normal play' will be a steady but lowered level of losses, while the chances of a big win on a winning streak are increased. The logic for this approach is, of course, equally flawed.

On the whole, the Labouchere can be an enjoyable system to experiment with, but, as with all similar betting systems, it is crucial to understand that the risk has merely been redistributed, not eliminated.

Optimal Betting

Many gambling strategies focus on the question of **what** to bet on. But equally important is **how much** to stake on each individual bet. We've seen the danger of trusting betting systems such as the Martingale in this respect. We've also mentioned gambler's ruin, the name for the common mistake of failing to scale your bets down when you are on a losing streak. However, it is also important to bet differential amounts depending on how much value there is in specific bets.

There are a few simplistic strategies you could use to regulate your stakes. One is to use **fixed stakes**, meaning that you gamble the same amount on each bet. The main advantage of this is that it prevents you from betting wildly variable amounts depending on your hunch or whim, and keeps your betting disciplined.

The disadvantage is that it fails to avoid the problem of gambler's ruin, so an obvious adaptation is percentage betting, whereby you bet a **fixed percentage** of your current bankroll on each bet. This is a slight improvement, but fails to allow for the variations in value you might find in games or sports more complex than a game like roulette.

One way you might address this problem is **fixed-return** betting. This means you identify how much you want to win on each bet (say, 1% of your current bankroll) and bet accordingly. This would mean that on a horse race you would bet 1% of your bankroll on an evens (2.00) bet but only 0.2% on a 5/1 (6.00) shot. This has the advantage of leaving you less exposed on the longer-odds bets.

I have seen gamblers argue against this strategy on the basis that you should always be betting on an outcome that you believe is probable and that fixed-return betting thus cheats you out of bigger returns on long-odds victories. However, it is better to think about bets in terms of actual probability versus implied probability – if you think a 5/1 horse has been underpriced by the bookmaker, it is still less likely to actually win than an evens bet that the bookmaker has underpriced would be, and so you should still take less of a risk on that bet.

Fixed-return betting is not the worst strategy in the world but, as we shall see, it can be improved on by using the Kelly criterion, which is a more precise way of targeting your stakes. However, before we discuss that, it is worth mentioning a couple of other approaches to optimal betting.

Due-column betting is similar to fixed-return betting, but significantly more dangerous. This is a system where you decide what proportion of your bankroll you want to win, and then bet enough on each individual bet to ensure that outcome if your bet is successful.

For instance, let's imagine you have $500 and have decided you want to win $20. You draw up a table like Table 2 and enter each bet until you have won this amount. The 'due' column starts out at $20 and increases until you have a winning bet, at which point you have made your target amount.

Bet number	Due	Odds	Bet	Result
1	$20	5/1 (6.00)	$4	Loss
2	$24	3/1 (4.00)	$8	Loss
3	$32	8/5 (2.60)	$20	Loss
4	$52	2/1 (3.00)	$26	Win

Table 2. Due-column betting

It should be fairly obvious that this strategy resembles the Martingale system in that it promises a set win in return for rapidly escalating stakes, and that a longish run of losses will be sufficient to wipe out the gambler. The idea that you are 'due' a win after a certain number of losses is known as the **gambler's fallacy**, a particularly dangerous piece of irrationality, which we will return to shortly.

Finally, there are situations in some games where varying your betting according to the game situation is advisable. In blackjack, if you know for sure that the ratio of ten value cards to other cards is relatively high, then it is advisable to increase your stake accordingly. This is the basis of cardcounting (see Chapter 5). In poker, it is common to use the ratio between pot odds (the ratio between a potential win and the bet you would need to make to call the other players) and true odds (the probability of receiving the card you would need to win) to judge whether or not to continue in a round of betting. These are both good strategies, but strictly speaking they aren't optimal betting: it makes more sense to view them as value betting opportunities since they are based on a genuine assessment of the gambler's edge as opposed to the implied probability of the odds on offer.

None of the pure optimal betting strategies we've looked at thus far is particularly satisfying. However, there is one strategy that stands out from the crowd as a genuinely useful method of assessing how much to bet: the Kelly criterion.

Kelly Betting: the Basics

Developed by J. L. Kelly in 1956, the Kelly criterion is a mathematical formula used to determine how much of your total bankroll to risk on a given bet. In many gambling and investment scenarios, it will give the gambler or investor the best chance of maximizing their profit, although only in situations where the house edge can be overcome.

It was specifically developed to work for a series of bets each of which carries equal odds and an edge which is in favour of the gambler. While many gambling scenarios don't provide this exact set-up, you can adapt the criterion for other circumstances. And, more importantly, if you take the time to study the method and to think about how it works this is a good way to ingrain common-sense habits about a sensible ratio of risk to stake.

In order to use the Kelly criterion, we first need a betting opportunity for which you have a positive edge and for which you have a realistic estimate of how big the edge is. The formula for how much of your bankroll to risk is:

$$\frac{bp - q}{b}$$

where

b is the decimal odds minus 1

p is the probability of success

q is the probability of failure (= 1 − p)

Imagine playing a game where you draw balls from a sack and bet on whether a red or black ball will come out next. The odds offered are evens (= 2.00 in decimal odds), but the referee is a friend of yours and they have accidentally let slip that there are actually 53 red balls and 47 black, so you have a $\frac{53}{47}$ edge when you bet on red.

In this case b = 1 so the Kelly criterion would suggest you bet

$$\frac{0.53 - 0.47}{1} = 0.06$$

So you should bet 6% of your bankroll each time in order to optimize your returns. This is not a guaranteed betting system and its success will still depend on you finding value betting opportunities: it is simply the best way to estimate how much of your bankroll you can place on each bet in order to maximize your return while minimizing the downside risk.

Remember that you also need to adjust your stake up and down to reflect your current bankroll as you progress. It is easy to set up a spreadsheet or other way of quickly entering the necessary data in order to apply the formula in fast-moving situations. And you can find a variety of online Kelly calculators if you want a shortcut.

One problem with the Kelly formula is the scarcity of betting opportunities that exactly fit the set-up requirements. Another is the difficulty of exactly calculating your edge in any given scenario. So mostly, when you use the formula, you will have to work with rough estimates of the correct figures involved. For this reason many gamblers work to a fractional Kelly system, for instance they might do the standard calculation, but then reduce their stake to a half or a third of the recommended one. This will reduce potential profits, but also reduce risk, so is a sensible strategy for using the formula where you aren't entirely confident in your estimate of the edge.

Either way, it is worth familiarizing yourself with the workings of the formula. It is pretty much the only pure betting system that has a long track record of helping people to make money without bankrupting them (as systems like the Fibonacci strategy are prone to doing).

And after a while working with the formula, you also develop a more intuitive understanding of how and why you should put different weight on different bets.

As we shall see later, it is also a tool that has served many investors well, including Warren Buffett and Bill Gross. They use it as a pure mathematical rule that helps them to allocate risk across their portfolio in the most effective way.

What to Do

At the very least, it is worth running some experiments in which you use the Kelly criterion for a series of bets. This is a useful exercise which can help you to acquire a greater understanding of how adjusting your stake according to the situation and the size of your bankroll can improve your chances of making a profit.

Hedging Your Bets

In the field of gambling, arbitrage and hedging are both names for the practice of betting for and against the same thing in such a way that you guarantee a winning margin or reduce your risk – like many moneymaking opportunities, both rely on variations in value.

Arbitrage generally refers to placing bets with different bookmakers so as to take advantage of differential odds, while hedging is more often used by gamblers to refer to a situation where you take advantage of changes in the odds over time. (It was the practice of taking up long and short positions on the same stock in this way that gave the name to 'hedge funds'.) I'll explain the basic principles of hedging, but bear in mind that the same maths can also be applied to arbitrage opportunities.

For instance, let's say a bookmaker takes your bet of $100 at 4/5 on (digital odds of 1.80) on the New York Giants to beat the Denver Broncos at the Super Bowl, and at the same time he is also offering bets on the Broncos at the same price, giving him an edge of 20% in total.[*] By doing this he is essentially selling you a promise to pay out $180 if the Giants win.

But the price of such promises can change due to circumstances (or another bookmaker may already be offering different odds). Maybe the Broncos star quarterback gets injured in the warm-up, or the Giants take an early lead during the game. Now you can get the Broncos at odds of 3/2 (digital odds 2.50). So you can nail down a guaranteed profit, whichever team wins, if you get your sums right.

The simplest way to work this out is to work back from the winnings you would get from the Giants bet, and divide by the digital odds on the Broncos:

$$\frac{\$180}{2.5} = \$72$$

---

[*] Assuming that there is no third option of a draw or tie.

So a bet of $72 on the Broncos will get you a win of $180 whoever wins. And your total cost will be $172, meaning you have ensured a win of $8.00.

This is, of course, not a huge win, but it is guaranteed.

If you expect the Giants to convert their advantage into a win, an alternative strategy is simply to cover your potential losses in case you are wrong. To work this out you need to know how much to bet on the Broncos to guarantee breaking even in total. This involves a tiny bit of algebra:

$$2.5x = \$100 + x$$

Subtract x from both sides:

$$1.5x = \$100$$
$$x = \$66.66$$

Or, if you want a shorter version, subtract 1 from the digital odds and divide your original stake by the result.

This will guarantee a larger amount, $13.34, if the Giants win, while making sure you lose nothing if the Broncos prove you wrong and win.

If you want to adapt hedging and arbitrage for situations with more than two possible outcomes, the basic rule is to start from the amount you want to win, then for each outcome divide by the digital odds and add the results to find your total stake.

As an example, consider a soccer match with the following starting odds:

Reds United to Win 1/5 (1.2)
Draw 5/1 (6.00)
Blues City to Win 17/2 (9.5)

If we were to bet on all three options at this level, with the aim of a \$10 payout, we would need to stake

$$\frac{10}{1.2} + \frac{10}{6} + \frac{10}{9.5}$$

which, with a bit of rounding up, comes to:

8.33 + 1.66 + 1.05 = 11.04

So the bookmaker has an edge of about 9%, which is fairly standard, and betting on all three outcomes would only guarantee a loss of £1.04. However, 15 minutes into the game there have been no goals, the Blues are looking better than expected, and the odds have shifted thus:

Reds United to Win 4/6 (1.66)
Draw 11/4 (3.75)
Blues City to Win 10/3 (4.33)

The bookmaker still has a similar edge on the spread of three outcomes: but let's say we had placed a bet on Blues City at the original odds, paying \$1.05 for a \$10.00 win.

Now we can take out covering bets on Reds United and the Draw as below:

$$\frac{10}{1.66} + \frac{10}{3.75} = 6 + 2.66 = 8.66$$

So added to our stake on the Blues we can pay 8.66 + 1.05 = $9.71 and get a payout of $10.00, and a profit of $0.29 whatever the outcome.

(A slightly weaker version of hedging is **dutching**, where you bet on several selections at once, aiming for a fixed return, without covering every possible outcome. The rationale is that you rule out some of the most unlikely outcomes, for instance the weakest few horses in a race, and then 'guarantee' a return if one of your chosen horses succeeds. Obviously this strategy can fail badly if your estimation of which outcomes to rule out is wrong.)

There are two main downsides of hedging. Firstly, in a betting scenario where the house has an edge, you can't guarantee that the odds will swing in your favour so as to allow a hedging bet. For instance in the first scenario above, if it had been the Broncos that strengthened in the betting, the opportunity to hedge would not have arisen. Similarly if the Reds had scored immediately after we laid our first bet in the soccer game, we wouldn't have been able to hedge successfully.

For this reason hedging is best applied in scenarios that have many changes in momentum. So high-scoring sports like tennis, in which the momentum can swing several times between the players, are more susceptible to in-game hedging than a sport like soccer, which might be settled by a single goal, creating one decisive shift in momentum. Meanwhile gamblers looking for arbitrage opportunities (often known as 'arbers') tend to focus on obscure sports or fixtures where there is more likely to be a variation between different bookmakers. (A warning: bookmakers hate arbers and keep a close eye out for telltale signs such as suspiciously precise amounts being gambled on Albanian handball games, so a degree of common sense is required to avoid having your accounts suspended. I couldn't possibly suggest that if you search the internet for gambler's forums and 'arbing' you will find plenty of less legitimate advice on how to avoid being noticed.)

The other drawback of hedging in gambling is that it can take a large stake to guarantee a fairly small profit. So the losses from one bet which doesn't provide a hedging opportunity can wipe out many smaller successful bets. It's worth bearing in mind that bookmakers have increasingly allowed in-game betting precisely because it encourages hedging (or cashing out winnings early, which is a mathematically identical strategy). The bookmaker's rationale is that while some people will profit through hedging their bets, others will miss out on bigger payouts by taking the smaller guarantee of a hedge. So it gives the gamblers a false sense of confidence: on the whole the bookmakers continue to profit from their edge, and the gamblers continue to lose.

However, it is worth understanding the mathematics of hedging as there are scenarios in gambling (and even more so in investment – see p. 135) where it can be a useful tool. Bookmakers have always safeguarded their own business model through a version of hedging, where they lay off bets with other bookmakers. This works because they always have the edge, but can nonetheless be exposed to big losses when a particular favourite wins – to compensate for this, they place bets elsewhere that will reduce their exposure on particular outcomes so as to smooth out their overall cashflow. This should serve as a reminder that, in a situation where the edge is in the gambler's or investor's favour, hedging is an excellent way to spread risk.

Hedging in Business

A good understanding of the maths of hedging can also be helpful in business scenarios. For instance, imagine an exporter in America who takes a large order for plastic bananas to be delivered to Germany – they are paying the manufacturer in dollars but being paid in euros. Their profit margin is 15% on the deal, but a swing in the wrong direction in exchange rates could easily take this down closer to zero. In this situation, the exporter is effectively taking a gamble on the exchange rates (bearing in mind that a change in their favour could increase their margin). Assuming the exporter has a euro account, one hedging option would be to sell euros to the value of the order at current exchange rates. This allows them to nail down the dollar profit margin at current exchange rates. Just as with a hedged bet, this means foregoing possible larger profits to avoid the risk of losses.

Value Betting

Whatever your betting strategy, whether you use the Kelly criterion, hedging or some other method of trying to preserve your bankroll and avoid ruin, the only medium-to long-term way to turn a profit from gambling is if you can genuinely find bets that are good value, meaning that the true probability is better than the implied probability of the odds on offer for a bet.

In Chapter 5 we will look at some of the less legitimate ways of trying to beat the system, from cardcounting to insider trading. At this point, however, let's stick to some of the legitimate ways you might find bets that are better value.

We saw in a previous section that some casino games can provide value bets where, for instance, the true odds are better than the pot odds in poker. But for the most part, casino games are so thoroughly stacked in favour of the house you will only find true value bets through high levels of skill in a game such as poker (especially if you're not playing the house), or through dubious means such as cardcounting.

In the field of sports betting, however, it is theoretically possible to use statistical analysis to outthink the bookmakers. Michael Lewis' book *Moneyball* is a fascinating look at how the received wisdom on what really influences a sports match can be wrong – it focuses on Billy Beane, the general manager of the cashstrapped baseball team the Oakland A's. He couldn't afford to buy the big name hitters and pitchers that would normally be seen as the path to success. Instead his staff compiled huge amounts of statistical data, which led him to focus on more obscure targets – for instance hitters with a high on-base percentage (a statistic that measures how often a batter reaches base). This allowed him to build a successful team for modest amounts of money.

British businessman Matthew Benham has made a fortune from statistical analysis, running a company that grew out of his fascination with gambling. He has achieved similar success at Danish soccer club Midtjylland, and is currently bringing the model to bear on the soccer club Brentford in England, where he has been a major investor. (He objects to the use of the term Moneyball to refer to his work at these clubs, feeling it has been misinterpreted to mean the generic use of statistics in sports management, whereas he sees it as specifically being about the scientific use of statistics.)

Whether you are applying statistics to sports management or to gambling, you need both a forensic approach to the figures and to have a method or system that is not already widely known. A good example of the reason why is the Beyer Speed Figures. This is a way of rating the performances of thoroughbred racehorses developed in the early 1970s (and published in the book *Picking Winners*) by Andrew Beyer, who wrote a racing column for the *Washington Post*. Given a past performance by any given horse, the number is based on the winning time, the time of the race and an allowance for the 'par speed' of the track over which it was run.

When these were first used by gamblers, they genuinely conveyed an advantage as they were a significant step forwards in the statistical analysis of results. However, over time, they came to be used as much by handicappers and bookmakers as by the punters, and thus no longer bestowed any significant advantage on the bettor.

The moral is that you should trust the statistics, but you always need to find new ways to analyse results in order to identify the games and moments within games that give you value opportunities. For instance in the book *The Numbers Game: Why Everything You Know About Football is Wrong*, authors Chris Anderson and David Sally point towards a range of traditional assumptions about the patterns underlying soccer games that turn out to be inaccurate on closer statistical analysis. For instance they observe that the worst player in a team more often has a decisive influence on the final results than the best player. And they also suggest that gamblers on soccer games tend to overrate how likely it is for corners to lead to goals – as a result the in-game odds tend to shift disproportionately when a corner is awarded, creating a good opportunity to bet the other way (since there is only one goal for every 45 or so corners).

David Sumpter, author of *Soccermatics*, has written about a brief experiment he conducted to try to find a winning model for betting on soccer. He tried a few different methods, such as modelling outcomes through expected goals based on the number of shots a team averaged from inside and outside the box. In the end he had modest success with two strategies. Firstly it is often the case that favourites in sporting events are overpriced by bookmakers, because gamblers have a bias towards longer-odds bets and the promise of greater returns. As a result the true probability on the favourites can sometimes be greater than the implied probability. Betting consistently on the strongest three teams in the league on this assumption would have brought a small profit in the 2014–15 English Premiership.

Sumpter also modelled a strategy based on the assumption that, because gamblers don't like betting on draws in soccer, the draw will often offer a value bet. He observed that this strategy would have won a fairly strong profit in the seasons he was analysing. (However, Sumpter's final gambit of simply asking his wife to guess the scores based on a quick look at recent results was the most successful of all the strategies he considered...)

These kinds of observations about statistics and betting patterns can genuinely be valuable, but there is also a danger that they can lead to overconfidence. For instance, while Sumpter's strategy of betting on the strongest three teams showed a success rate in 2014–15, the same strategy would have failed in the following season, when there was a greater proportion of games in which the underdogs overturned the odds. (2015–16 was the season in which Leicester City went from being 5,000–1 outsiders at the outset to becoming champions.)

So any attempt to use statistical analysis requires that the gambler have a large enough sample on which to base their conclusions and a good familiarity with some of the rational fallacies that can undermine the smartest minds in both gambling and business circles.

The Laws of Small and Large Numbers

We tend to be far less statistically savvy than we think we are – where maths is used to analyse sports, casino games or markets we should never rely on insufficient data or use the maths tools simply as a prop for our hunches. If anything, we need to use maths as dispassionately as possible.

There are several laws and fallacies that tend to affect our judgement about statistics and probability. Firstly, the **law of large numbers** is the principle that a very large sample of a set of outcomes is likely to be close to the probable or average value of all members of that set. For instance, if the average height of adult men in Happyland is 5 feet 7 inches, then the larger our sample, the closer the average height will be to this figure. A very small sample might be disproportionately biased by a few tall men and give us an average of 6 feet. A slightly larger sample will tend to be closer to the mean and, as the sample increases, the average will converge towards the mean of the population.

Figure 18. Large and small samples.

The law of large numbers is the reason why casinos can be sure that they have a profitable model. If you run one roulette wheel for an hour, you can lose a significant amount of money in spite of the house edge. If, however, you run 50 roulette tables for a month, the sheer quantity of games will mean that the winnings will be close to the figure predicted by the house edge.

By contrast, the **law of small numbers** is the principle that we often draw erroneous conclusions from small samples. For instance, if a new salesperson working for your company has three good months in a row, then you might conclude they are a brilliant salesperson. But this could simply be a result of short-term volatility in the results, and the following nine months may be poor ones. Or the salesperson may simply have stumbled on a run of luck in a new territory, or have benefited from reorders of the previous salesperson's work, which would mean that we weren't even basing our judgement on the right information.

The law of small numbers was formulated by Daniel Kahneman and Amos Tversky, who made numerous significant contributions to our understanding of the psychology of decision-making. Their most significant breakthrough was the simple observation that people's hunches about probability can often be wildly wrong. Previous research had tended to rely on the fallacy that where a large group of people make independent estimates about the probability of outcomes in a particular situation, they will be distributed evenly around the correct probability – in other words, some people will guess too low, some will guess too high, and the mean of their guesses will be more or less correct.

Psychologists Tversky and Kahneman ran many experiments which revealed how false this assumption was, and that some of the ways in which people get probability wrong are quite predictable if you drop this assumption. Basically, we're all much more stupid about probability than we think.

For instance they identified the **availability bias**, which means that people are prone to judging probability based on the most recent similar example they can recall. Conmen and salespeople take advantage of the availability bias by, for instance, reminding potential customers of profitable ventures and successful purchases before attempting to make a similar sale.

Tversky and Kahneman also developed prospect theory, which was an attempt to systematically lay out some of the rational biases they observed. A couple of examples from their work demonstrate how the **framing** of a question can affect how the decision-maker responds. First, consider these two options:

Scenario 1: You start with $1,000. You can choose between:

A.Winning $1,000 or $0 with a 50% probability of each

B. Getting another $500 for sure.

Scenario 2: You start with $2,000. You can choose
between:

C. Losing $1,000 or $0 with a 50%
probability of each

D. Losing $500 for sure.

In each scenario you are being asked to choose between a definite $1,500 and a 50% chance of ending up with $1,000 or $2,000. When these scenarios are tested experimentally, more people tend to be risk averse in scenario 1 (a higher percentage take the sure $500 over the chance of a bigger gain). However, in scenario 2, a higher proportion of people are risk-seeking and risk losing $1,000 in the hope of avoiding loss altogether.

There are also issues of framing at work when we find that most people will drive for ten minutes to save £10 on a £50 toy, but will not drive ten minutes to save £20 on a £20,000 car. In this case their decision is being affected by how large a percentage of their purchase the saving is, so again this is an example of our rational judgement being affected by how the problem has been framed.

What to Do

Do bear in mind how much your time is worth – try to separate this issue from the context. It is, for instance, easy to spend far too much time on work projects which have a low potential return (such as attending endless meetings), and too little on those with a potential high payout (such as prospecting for potential customers). It is sensible at all times to have a notional hourly rate in mind against which you can measure the potential savings and gains of any decision you make about business or investment. And try to overcome your natural biases such as framing and the availability bias to make a clear rational assessment of your actual chances of winning or losing from any given gamble or venture.

The St Petersburg Lottery

Imagine a game where you had to choose how much to pay for a ticket. A coin is tossed repeatedly, and your prize depends on how many times it falls on tails before the first head. If the first head is on the first toss, you win £1. If it is on the second, you win £2, if it is on the third, you win £4, if it is on the fourth you win £8 and so on, with the prize doubling each time.

The question is how much you should pay for a ticket

to this lottery. Given what we have discussed so far in this chapter, your first thought might be to calculate the expected return of a ticket. That would be a 1 in 2 chance of £1, a 1 in 4 chance of £2, a 1 in 8 chance of £4 and so on.

If we add up this series to find the expected value we get:

$$\left(\frac{1}{2} \times 1\right) + \left(\frac{1}{4} \times 2\right) + \left(\frac{1}{8} \times 4\right) + \dots = 0.5 + 0.5 + 0.5 + \dots$$

So the expected return on a ticket is infinite, and any finite price, no matter how large, would theoretically be a rational price to pay. However, most people wouldn't pay more than a modest proportion of a week's salary for a ticket. So what has gone wrong with the expected value model here?

This paradox was invented by the eighteenth-century mathematician Nicolaus Bernoulli, and its first resolution came from his brother Daniel. He noted that 'the determination of the value of an item must not be based on the price, but rather on the utility it yields... There is no doubt that a gain of one thousand ducats is more significant to the pauper than to a rich man though both gain the same amount.'

The concept of utility is now fundamental to economics. It explains why different people value the same item differently: because they will get a different level of satisfaction from using the item. Also implicit in Bernoulli's words is the concept of the **diminishing marginal utility of money**. Marginal utility is the satisfaction you receive from consuming an extra unit of a good. And diminishing marginal utility simply means that the more you have of any given good, the less satisfaction it will give you. At a basic level, one slice of cake might be delicious and filling. A second slice might still be satisfying. But by the third or fourth slice your pleasure in eating the cake will be significantly less than it was for the first slice.

Most goods have diminishing marginal utility. And money, which can be treated as a good in its own right, also has diminishing marginal utility. A five ducat coin is worth a great deal more to a pauper than it is to a duke. So while the duke might casually gamble that coin on a game like the St Petersburg lottery, the pauper would need to hang on to it to pay for food and shelter.

Bernoulli suggested a formal definition of utility as being proportional to the wealth of the gambler, and derived the scary equation below (which I'm not even going to pretend to be able to explain) as a way to establish the value of the lottery ticket (where c is the price of the ticket and w is the wealth of the gambler).

$$\Delta E(U) = \sum_{k=1}^{\infty} \frac{1}{2^k} [\ln(w + 2^k - c) - \ln(w)] < \infty$$

There have been plenty of other attempts to provide an ideal mathematical function to define utility, some more comprehensible than others. But the crucial concepts that we need to remember don't rely on such complex equations. They are simply that (1) it can be rational for people to place a different value on the same goods because they will base their estimate on utility; (2) utility is subjective, relative to the wealth of the consumer; (3) money itself has diminishing marginal utility.

Another Kahneman and Tversky question was this: why do a majority of people choose not to gamble $50 for a 60% chance of winning $50. To understand this, we need to think about the financial position of each individual who is asked the question. If they have enough money to regard $50 as a minor loss, they may well find it rational to take this gamble. However, if losing $50 will mean they are out of cash until payday, it clearly makes more sense to decline the gamble. The money itself has a different level of utility in the two situations. And for each gambler the first $50 (the stake) has more utility than the second (the potential winnings) – it's just that for the poorer individual there is a sharper distinction between the money they can't afford to lose and the money they could potentially win.

Utility in Business

In business in general, it is crucial to understand diminishing marginal utility when considering business models and new product lines. A business is only sustainable if it can keep selling a product over time, and this generally means that it needs customers who make repeat purchases.

Some products, such as food, are destroyed by being consumed, so naturally require regular top-up purchases (although to succeed with a specific food product you also need to encourage habitual use of that product). The idea of planned obsolescence is understood by many businesses, who intentionally produce products that require regular replacements or upgrades. Of course planned obsolescence should be a worry to anyone concerned about the environment – but it is a very hard thing to avoid in our economic model.

Until the 1970s there used to be numerous non-disposable safety razors on the market – these could last a lifetime and, in some cases, you could sharpen the blade rather than replace it. However, this was not in the long run a sustainable business model and most companies switched over to the disposable plastic razor as a result. In other cases, the market can simply become saturated with a particular product.

Recently there was a huge boom in colouring books for grown-ups. The first few successes in this area were followed by a flood of beautiful and imaginative colouring books from publishers around the world. The problem was that many grown-up fans of colouring soon reached a point where they had enough colouring books to last a lifetime and consequently stopped buying new ones. As a result the market for these books crashed, and many publishers ended up having to pulp their most recent titles. The moral is that you need to understand how rapidly the marginal utility of your product will decline, and have contingency plans for dealing with this problem.

Practical Probability and More Irrational Traps

We have already mentioned gambler's ruin (failing to adjust your individual stakes down in proportion to a reduced bankroll) as an example of the kind of irrational beliefs and practices that can hinder gamblers. Another related problem is **gambler's conceit**, which is when a gambler believes that they will be able to stop engaging in a risky behaviour at exactly the right moment. In practice this means that many gamblers find it hard to quit while they are ahead. Gambler's conceit is rooted in a basic human cognitive bias: the **optimism bias.** This is the tendency to believe that we are luckier than other people, or more likely to succeed in a particular risky endeavour. It is a difficult bias to get away from, which is why it is so important to understand that in the long run all lucky runs will tend to dissipate as the law of large numbers means our results tend closer to the mean result, which will be a loss of the expected value where house edge is in play.

It is also important to understand the **gambler's fallacy**. This is the tendency to expect that, where there has been a run of results in a game that is biased in one direction, this makes it more likely that ensuing results will be biased in the opposite direction. For instance, if we watch someone toss a coin five times and it always lands on heads, the gambler's fallacy would be to believe that it is now more likely to fall on tails.

This is also known as the **Monte Carlo fallacy** because of events at the Monte Carlo Casino on 18 August 1913. At the roulette table, the ball fell on black a freakish 26 times in a row (with a probability of 0.0000023). Some gamblers might have taken the view that this could be an unfair table, but most of the gamblers at the casino at the time instead believed that a long streak of red must be due, and lost millions of francs betting on this outcome. In fact, starting from the point directly after the last of the 26 consecutive blacks and assuming a fair table, the true probability of every ensuing roll of the ball was equal for red and black.

It is important to be careful about the distinction between the gambler's fallacy and **regression to the mean**.

This term refers to the fact that if a variable is extreme (in other words unusually far above or below the mean) on its first measurement, it will tend to be closer to the mean on the next measurement. The concept was first identified by the geneticist Francis Galton in the nineteenth century – he noticed that unusually tall individuals tended to have children who were less tall than themselves. Where there is any element of randomness in a process, we tend to see a similar pattern. For instance if you measure the number of cars jumping red lights at different locations in a city one day, then repeat the measure the next day, you will tend to find that the locations with the highest results on day 1 will have slightly lower results on day 2, while the ones with the lowest results on day 1 will have slightly higher results on day 2. The overall distribution will not be narrower as a result, since other locations will now be taking the outlying points.

It is important to understand that this is a **result** of randomness, not an exception to it. If you drop balls through a bean machine like the one shown in Figure 19, the balls can bounce either right or left at each pin and will tend to end up in the channels below with a normal distribution. (A bean machine or quincunx is a kind of bagatelle device with pins that the balls bounce on as they fall.) Suppose you then drop the resulting set of balls from their end positions through a second bean machine, after which they will again end up in channels with a normal distribution. If you ask where the balls that now lie at the extremes of the distribution, beneath the outliers from the previous distribution, are most likely to have fallen from, the correct answer is that they are more likely to have fallen from closer to the centre than from directly above. This is simply because there are more balls that start out in the channels closer to the centre, so there is a much higher chance of these balls bouncing out to the wider locations.

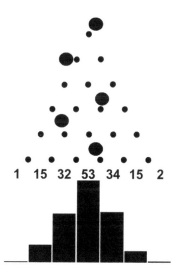

1 15 32 53 34 15 2

Figure 19. Each ball is dropped from directly above the centre of the quincunx and then ricochets down the pegs, creating an approximation of the normal distribution in the final destinations of the balls.

As another example, in a soccer league, if a team has an unusually good start to a season and wins five of its first six matches, it is a mistake to expect the final win proportion to be $\frac{5}{6}$ – it is more likely to be lower than this, rather than higher. Of course it may simply be that the team is in excellent form that will continue all season – but it makes sense to at least consider whether regression to the mean is likely to occur before staking too much money on this rate of success continuing.

Here's one way to understand the difference between the gambler's fallacy and regression to the mean. If we have a competition where 1,000 people toss coins, and those who get heads rather than tails stay in the competition at each round, we would expect roughly 500 to qualify from the first round, 250 from the second, 125 from the third, 62 from the fourth (rounding down) and 31 from the fifth.

At this stage we have roughly 31 people who have thrown five heads in a row. The gambler's fallacy would be to look at any one of those players and predict that their next throw is more likely than 50% to fall on tails, on the basis that we are expecting the heads and tails to 'even up'. Of course the odds for each of those players to throw a heads or tails on the next throw remains at 50% and, measured from this point, we would still expect their future throws to average out at 50% heads and tails (meaning that our prediction **from** this point is that their mean throws will be distributed around a total of five more heads than tails).

If instead we focus on one of these players and rerun the experiment, it is rational to expect that their next attempt at the game will fall nearer to the mean and they are unlikely to throw another five heads in a row. This is because of regression to the mean.

The Search for Patterns in Randomness

Humans aren't very good at understanding randomness. As part of the fundamental way we understand life, we have a tendency to look for patterns, and this can lead to us seeing patterns that aren't really there. It also means that when we are asked to simulate randomness we do a fairly bad job of it.

Here are a couple of examples that should clarify this point. In the 1890s, the Monte Carlo roulette results used to be published daily in the newspaper *Le Monaco*. The mathematician Karl Pearson wanted to test some of his methods on random data, and utilized the roulette results as recorded in the newspaper for this purpose. His conclusion was that the roulette wheels were behaving extremely oddly. The types of runs of results they showed were not random at all and gave the appearance of being rigged or biased to an extreme degree. In fact, the lazy journalists at the newspaper had decided that no-one would notice if they simply made up the results, so rather than discovering that the roulette wheels weren't random and the casino was rigged, Pearson had merely demonstrated how bad the journalists were at mimicking random data sets.

If we are asked to imagine a series of random roulette wheel spins, our naive view of how a random process will look leads us to generate series of results that are consistently erratic, that don't contain anything that 'looks like a pattern' and that spread the results around fairly evenly. In fact a genuine random process can look quite clumpy, can have series of results that look very 'pattern-like', and, for instance, might not land on the number 7 for hundreds of spins, in spite of landing frequently on the number 6. Genuinely random events often don't look random to us.

A related problem afflicted Apple when it introduced the shuffle feature on the iPod. Many users noticed songs by the same band occurring close together, or the same song recurring, and complained, wrongly assuming that the shuffle was somehow rigged rather than being truly random. Steve Jobs was offended by the mathematical slur, but nonetheless amended the product, explaining that they had had to make the shuffle 'less random to make it feel more random'.

So our perceptions of randomness are truly unreliable. One result of this is gamblers' tendency to perceive 'streaks'. It is a common superstition that online gambling software is unusually 'streaky' and therefore not random – in fact the occurrence of streaks of results is a fairly normal occurrence in a random process, so this is almost always a result of our faulty perceptions rather than genuine bugs in the software.

Our failure to understand randomness leads to two more rational traps. Firstly, one version of the gambler's fallacy is the tendency to leap to conclusions about streaks based on fairly limited data. We might for instance see that a football team has drawn three games in a row and is therefore more likely to draw the next one. Or that a company has had ten weeks of growth in its share price, and wrongly conclude that this streak will continue. Within the context of sports betting, this is known as the **hot-hand fallacy**, a name that comes from a 1985 paper by Thomas Gilovich, Robert Vallone and Amos Tversky. This studied the apparent misperception that a basketball player is more likely to score from a shot when his previous shot was successful. In a sense this is the opposite of the gamblers' fallacy as it relies on the idea that an unusual streak of results will continue rather than assuming that luck will 'even out'.

It is important to be aware of such fallacies, but also to be careful how we apply that awareness. There is some evidence that sportspeople do gain confidence from success and will therefore genuinely be more likely to score following a recent success. There is even evidence[*] that gamblers can have lucky streaks – in this case it is for a somewhat ironic reason. Many gamblers tend to expect their own lucky streaks to end. So as they win a series of bets their optimism about the next bet can gradually fall, leading them to take on increasingly safe bets and thus become more likely to keep winning.

---

[*] http://www.popsci.com/article/science/are-lucky-streaks-real-science-says-yes

Finally, do bear in mind that sometimes an apparently random process really does have a bias. The story of Joseph Jagger, a mechanic from Yorkshire who had huge wins on the Monte Carlo roulette tables, might be apocryphal – but whether it is true or false there is a moral to be observed in it. Supposedly, Jagger hired a team of clerks to note down weeks' worth of results from the roulette wheels and identified some of them as being biased towards particular numbers due to mechanical imperfections or unevenness in their positioning. This apparently led to him successfully overcoming the house edge and walking away with a fortune.

The story might not be true, but casinos certainly believe this could happen – they are very careful about swapping roulette wheels around regularly to avoid future imitators. Either way, it demonstrates an important truth. To identify a genuine bias or streak, you would need a huge amount of data. If you can amass such a store of information, then you may be able to identify real patterns and biases in a sport or game. But if you attempt to do so on insufficient data, then the difficulty we have in understanding randomness is all too likely to lead you to erroneous conclusions.

What to Do

While the varying effects of the gambler's fallacy, the hot-hand fallacy and regression to the mean can be confusing in practice, the main things to remember are not to rely on apparent patterns in genuinely random events, and that past events do not affect the probability of genuinely independent future events.

Roulette and Formal Mathematics

The study of probability in roulette has had some surprising and useful side-effects in other fields. The mathematician Henri Poincaré (see p. 291) used a modified version of a roulette wheel to demonstrate how sensitivity to initial conditions can theoretically be used to predict the ultimate resting state of the wheel. However, as an aside to his work, by showing how tiny changes to those initial conditions could lead to major changes in the final result, he also laid down one of the foundations of modern chaos theory, which shows deterministic physical processes can lead to outcomes that are in practice virtually impossible to predict. (All that we really mean when we say that a dice throw or roll of the roulette wheel is 'random' is that the deterministic processes that define its outcome are too difficult to analyse in the time available.)

Meanwhile the methods Karl Pearson used to unveil the bluffing journalists of *Le Monaco* have become a foundation stone of modern science. When analysing test data for new medicines, or for experiments on physical or chemical processes, scientists test theories against the probability of obtaining this particular result through luck alone (see statistical significance, p. 235). This is the fundamental method used to check whether the evidence actually supports a theory or whether we are just being befuddled by randomness, and also to detect fraud and made-up results.

Chapter 3 Summary

1. We are all prone to irrational mistakes while gambling, and the best-laid plans can all too easily go wrong.
2. Strategies such as hedging and using the Kelly criterion can improve your chances of making a profit or limiting your losses.
3. There are many more irrational reasons to bet and betting strategies than rational ones. The only truly rational reason to bet is if you have some knowledge or analysis that allows you to overcome the house edge.
4. All foolproof or 'surefire' betting systems are bunkum, without exception.

CHAPTER 4

The Successful Investor

An investment in knowledge pays the best interest.
Benjamin Franklin

Since gambling is generally a bad moneymaking strategy, it makes sense to look for situations in life where the edge may be in your favour rather than working against you. The stock market is, in theory, such an opportunity, since it tends to show long-term real growth. This means that, in contrast to the casino, the investor will tend to have the edge. So while there is much that can go wrong, the playing field for an investor is more even than for a gambler (although the maths and data can be much harder to analyse). Let's take a look at the absolute basics of investing in the stock market, then move on to some of the other ways that maths can be important in a business environment.

First, a warning: there's a limit to how extensively we can unfold the various equations and calculations given in this chapter. My aim here is just to demystify the jargon, and to explain enough of the fundamentals to allow people to continue to explore areas that they find interesting and to have a basic common-sense grounding in the subjects covered in investment textbooks and courses.

The Stock Market: An Introduction

In some respects, buying stocks, bonds or shares is similar to betting. You are purchasing a financial instrument which you hope will be worth more than you paid for it in the short, medium or long term. If you are hoping for a short-term increase in the value, then you are engaging in speculation. If, however, you are willing to leave your money in the company for a longer period, you can legitimately say that you are investing in the company.

There are different types of financial instrument, which carry differing levels of risk. For instance, bonds, which are sold on the bond market, are traditionally a safe investment – they are essentially a way of lending a company money at an agreed rate of interest (but which won't increase in value with the company's performance or pay dividends).

Shares, which are sold on the stock market, do have the potential to grow in value and also to pay dividends, a share of the company's profits. However, dividends won't be paid if the company chooses to retain its profits and reinvest them in the company, in which case improved performance should be reflected in a higher share price.

Shares in a company are initially owned only by the people who start that company, and any external investors they may have relied on. Further investors may then inject cash into the company in return for further private shares. However, if there comes a point where further capital is required, or where the initial investors want to cash in on their stakes, the company may consider an initial public offering (IPO) in which shares are offered on the open market, meaning anyone can buy them, and the company goes from being private to public. Owning stock also gives you voting rights in shareholder meetings, and confers the right to sell your shares to somebody else.

Shares come with different levels of rights and risk. Preferred stock grants better voting rights than common stock, while some companies issue different tiers of share: for instance Berkshire Hathaway, Warren Buffett's company, has issued stock in Class A and Class B tiers.

If you are interested in taking an active role in overseeing the affairs of a company or even trying to take control of it, it is important to have stock that grants the best possible level of voting rights, as this allows you to have an influence on who is on the board and so on. However, for the more passive investor, the importance of different types of financial instrument rests on their varying levels of risk. If a company goes into liquidation, the holders of bonds and other debt instruments will be paid first, then the holders of preferred stock, and finally the holders of common stock (and different classes may similarly be distinguished).

The level of risk involved is directly connected to the

potential rewards. Bonds have traditionally paid roughly 5–7% annually (and are fixed in value, meaning that when they reach maturity, the company will pay back the same price regardless of how it has performed in the meantime), while shares have returned an average of 8–9% (and can, in theory, grow in value without limit). So bonds are the equivalent of a short-odds bet, with very little chance of being wiped out, while shares are the equivalent of a longer-odds bet, which come with a higher risk of losing all your money.

A Share of What?

So, if you own a share of the company, what do you actually own? You can't just turn up and help yourself to a box of paperclips from the stationery cupboard on the basis that it's partly yours. This is for a good reason, which is that personal property and corporate property are legally separated and, as a shareholder, you have limited liability (which also means that a judge can't order your property to be seized if the company fails).

However, a share does give you a share in the total value of the company, which is known as the **market capitalization** (or market cap). At the simplest level, if a company has 50,000 shares in the same class and each is worth $20, then the market capitalization of the company is $1,000,000 (= $20 × 50,000). This means it is pointless simply comparing the share prices of two companies – the information is only meaningful when combined with the number of shares issued.

The market capitalization is made up of two elements: the **book value** is based on all the stuff the company owns – its inventory, buildings, office furniture, cash holdings, patents and copyrights, and so on. This is relatively straightforward to value (as long we are dealing with an ordinary company and not something outlandish like Enron...) as it is equivalent to an individual's net worth – the amount of money that would be left over if they sold off all their assets and paid off all their debts at current values.

The more volatile bit of the market value is made up of the future cashflows of the company. The main reason that share prices rise and fall is because investors have varying opinions on the value of a company's future cashflows, and sentiment in the market as a whole changes with events.

As individual investors issue buy and sell instructions to their brokers, the total supply and demand in the market causes the price to rise and fall. The most rational way to value a company is by carefully analysing its book value and future earnings potential. However, investors use a wide variety of methods, both rational and irrational, to decide what to invest in, so it would be too simplistic to say that it is just the varying projections of future cashflows of the company that make its price fluctuate. For a start, many people will be influenced by recent changes in the share price – upward movements may indicate a company performing well, in which case it can be rational to expect the trend to continue, but there is always the danger that the correction in the price has already overshot and is only being sustained by irrational exuberance.

Other things that affect subjective and objective valuations include the value of its branding, its human capital (in terms of how well trained and established its staff are), and barriers to entry (in other words, how easily new entrants to the industry can compete with the company). And you also have to take into account the effects of random factors such as crazy internet rumours, intentional attempts to manipulate the price, good and bad PR for the company, and so on.

Obviously as an investor your best strategy is to buy stock that is undervalued and either hold it or sell it when you believe it is overvalued or wish to transfer your money to other investment opportunities. But there is no guaranteed, foolproof way to value a company or to tell whether its stock is overvalued or undervalued. So let's take a look at a few of the ways that companies can be analysed.

The Price-to-Earnings Ratio

One traditional tool which has been used for over a hundred years is **the price-to-earnings ratio** or P/E ratio. It can be used to consider whether the stock is underpriced or overpriced and how it compares to other companies in the same industry or to the wider market. It is calculated by dividing the market value of the company by the annual earnings.

Imagine Motormouth Inc., a company that makes model cars, which has been generating a profit (earnings) of $30,000 a year, and which has been valued for sale at a price of $600,000. The P/E ratio is $\frac{600,000}{30,000} = 20$. This tells us that it would take any buyer 20 years to recoup the proposed cost of the company (if earnings stay constant). We can also compare this to other model car manufacturers or to other toy companies for a sense of whether it represents fair value. External factors, such as the influence of a new movie about the adventures of a model car, or newspaper scare stories about the dangers of children choking on model cars, would affect other companies in the same sector equally, so this comparison helps to give us a more precise sense of what is particular about this company.

If you wanted to buy stock in a particular sector, you might choose to use this yardstick to compare companies and invest in those with the lowest P/E ratios. It is generally reckoned that 20 is a fairly average P/E ratio while a lower ratio of say 10 might be seen as a signal to buy. Bear in mind that new companies tend to trade at a higher P/E ratio than mature companies, as the former can claim to have a less restricted growth horizon (meaning that it is harder for a long established company to find new, unexploited markets). You also need to have an understanding of this business model to interpret P/E ratios. For instance, Amazon, which continues to sacrifice short-term profits in favour of aggressive expansion into new markets, has recently had a P/E ratio of more than 500.

A related tool which can help with this kind of complication is the **PEG ratio**. This is the P/E ratio divided by the five-year earnings growth rate. For instance a company with a five-year growth rate of 30% and a P/E ratio of 15 would have a PEG ratio of 0.5. A PEG ratio under 1 is usually considered to indicate an undervalued stock. In a similar vein the **PEGY ratio**, which is often used by investors who prefer stocks that pay dividends, is calculated as P/E ratio / (growth rate + dividend yield). Again, a ratio of less than 1 is preferred, with ratios of less than 0.5 being thought to offer a good buying opportunity. If the forward P/E ratio (in other words the P/E ratio calculated using the coming year's projected earnings) is higher than the current P/E ratio, it means that analysts are expecting earnings to increase.

Of course the danger of projections based on current or past figures is that they may be flawed and that circumstances may change in unpredictable ways. A mathematical financial analysis tool is only as good as the information it is relying on: as one of my old bosses used to say about spreadsheets and databases, 'put garbage in, you'll get garbage out'.

What Not to Do

It's easy to feel overwhelmed by the sheer range of financial ratios out there, for instance the Gordon equation, which is a way of calculating expected returns from company dividends; as well as a whole range of other ratios relating to solvency, liquidity, efficiency and profitability as well as straightforward valuation ratios. The danger of getting too obsessed with these ratios is that you can get bogged down in the pure numbers and forget that these are all simply rules of thumb that are dim reflections of a complex set of facts and events in the real world. Mathematical interpretations can be extremely useful to an investor but it is best not to get too bogged down in them: experienced investors such as John Bogle and Peter Lynch stress the need to use a limited number of analysis tools that you are completely familiar with, alongside the imperative to invest only in companies that you truly understand, in order to avoid being misled by freak results or erroneous information.

The Time Value of Money

The mathematical logic of the P/E ratio becomes more intuitively obvious once you realize that this is a special case of a broader mathematical problem known as the **time**

value of money. This problem has been analysed since at least the fifteenth century, when it was discussed by the early Spanish economist and theologian Martín de Azpilcueta. It is the notion that a given amount of money theoretically has a greater value now than it will in the future, because the owner of the money should be able to earn interest in the meantime. For instance if I am offered $1,000, to be received in a year's time, how much should I pay for this opportunity. Let's say I could earn 5% by leaving my money in the bank, then we need to divide by 1.05:

$$\frac{\$1000}{1.05} = \$952.38$$

So to get that future $1,000 (assuming an interest rate of 5%) you would need to invest $952.38 now. In this equation the rate of 5% would be described as the **discount rate**, meaning the rate by which we have to discount the future value to find the present value. Of course the choice of interest rate is key: accountants often use the **risk-free interest rate**, which is the highest rate that can be guaranteed to be available from a bank or from government bonds. A more accurate but less cautious method is to use the **real interest rate**, which is the nominal risk-free interest rate minus inflation. (In other words if the risk-free interest rate is 7% over a period in which there was 4% inflation the real interest rate is 3%.)

However the rate is chosen, this is a demonstration of the basic fact that a rational person, given the availability of interest-bearing bank accounts (or other ways of using money to make more money) will choose $1,000 today over $1,000 at some specified future date. This raises the question of how to calculate the present value of money in general. The most fundamental equation for this is:

$$FV = PV \times \left(1 + \frac{i}{n}\right)^{(n \times t)}$$

where

FV is the future value of the money

PV is the present value of the money

i is the interest rate

n is the number of compounding periods per year

t is the number of years

If, for instance, we invest £20,000 at an annual rate of 6%, with the interest charged quarterly, we can calculate the value in two years' time as:

$$20,000 \times \left(1 + \frac{0.06}{4}\right)^{(4 \times 2)} = 20,000 \times (1.015)^8$$
$$= £22,529.85$$

And if we want to find the present value of a future payment, we can express the same equation as:

$$PV \times \left(\frac{FV}{\left(1 + \frac{i}{n}\right)^{(n \times t)}}\right)$$

There are different versions of this basic equation, depending on factors such as whether we are valuing an annuity (a regular payment), a perpetuity (a regular payment that will continue indefinitely), whether we expect the return on our money to increase over time steadily or exponentially, and whether the payments are due at the start of a period or the end of a period. These would all be useful at various times in the valuation and pricing of, for instance, leaseholds, mortgages and pensions.

For now, let's focus on one of the simpler examples, the formula for establishing the present value of a perpetuity. We need to calculate the sum of the present value of all the future payments. From a fairly complicated initial formulation, this simplifies down to this equation for a perpetuity with constant payments:

$$PV \times \frac{C}{i}$$

where i is the interest rate and C is the annual payment. We can rearrange this as

$$\frac{PV}{C} \times \frac{1}{i}$$

So present value divided by the annual payment is the inverse of the interest rate. It's fairly obvious that this is analogous to the P/E ratio, in which we divide current valuation by current earnings.

It thus makes a lot of sense to think of the P/E ratio as

being the inverse of the interest rate you are effectively being paid on your shares. If you were to buy a company at its current valuation, you would receive the earnings as payment for investing your money (and when you buy a share the same thing is happening pro rata) – so you would want this to be more lucrative than simply leaving your money in the bank.

This is why a P/E ratio of about 20 (the inverse of 5% interest) is seen as solid but unspectacular while a P/E ratio of about 10 (the inverse of 10% interest) is seen as a potential buying opportunity. Of course if you see an opportunity where the P/E ratio of a company is favourable, time may not be on your side: other investors will also see a chance to earn the equivalent of 10% or more interest on their money, there will be more buyers than sellers and this will push the price of the company upwards meaning that later buyers will not get such a good offer. But those who buy at the current price are still likely to be rewarded through growth in the value of their shares and future dividends.

Similarly we can now see why the PEG ratio is used as a yardstick. When you divide the P/E ratio by the five-year growth rate, you are simply comparing the effective interest rate available for investing your money with the actual rate at which earnings have been increasing. If the latter is higher and the PEG ratio is under 1, it is possible that the market has been underestimating the ability of the company to generate increased profits.

Six Out of Ten Ain't Bad...

The well-known investor and author Peter Lynch has said of picking stocks that 'if you're good, you're right six times out of ten. You're never going to be right nine times out of ten.' His point is that a portfolio which contains slightly more winners than losers is all you need to have a profitable investment. This is worth bearing in mind given how daunting the process of trying to identify good investments can seem. If you rely on a few well-understood mathematical yardsticks and do the research beyond that, you have a reasonable chance of success, given that the average stock tends to increase in value over time. (This rule relies on the idea of a stock portfolio in which the most you can lose on a given investment is the money you invested: when it comes to derivatives, which we will look at in Chapter 6, the picture becomes more complicated.)

☺☺☺☺☺☺☹☹☹

Managing Uncertainty in Decision-Making

The global financial crisis has demonstrated many things, including how badly things can go wrong for investors who put all of their eggs in one basket. One of the most important roles of mathematics in investing is to help spread risk in an uncertain world. It is impossible to avoid risk altogether, but there are numerous strategies investors can use to manage that uncertainty. In this section, we will take a very brief look at the essentials of **portfolio theory**, which is concerned with assessing the expected return of individual investments, and of a portfolio of investments as a whole. The main aim is to avoid correlated risk, in which all of the individual investments tend to move in the same direction in response to events. In order to do this, we are usually looking for **diversification** – this will generally lead to a reduction in risk so long as you avoid too much correlation across your portfolio.

To calculate the **expected return* for an individual investment** in portfolio theory, we find a way to calculate the range of possible outcomes and their probability. (Of course part of the art of financial investment is to find good ways to make these estimates in the first place.) For each possible outcome we multiply the return by its probability and then find the sum of the results.† For instance, say we chose to invest in Raaaah!, a company selling toy dinosaurs, and we have calculated the following probability table:

---

* Note that expected **return** in investment theory is equivalent to expected **value** in gambling parlance.

† If you prefer the algebra, the equation for the expected return, $E(R)$, is: $E(R) = p_1 R_1 + p_2 R_2 + \ldots + p_n R_n$, where p_n is the probability of a scenario and R_n is the expected return in that scenario.

Scenario	Probability	Expected return
Best case	20%	16%
Worst case	20%	-4%
Base case	60%	6%

Table 3. Probability table for Raaaah!

We calculate the expected return as:

$$(0.2 \times 16\%) + (0.6 \times 6\%) + (0.2 \times -4\%)$$
$$= 3.2\% + 3.6\% + (-0.8\%) = 6\%$$

So the expected return for Raaaah! is 6%.

Next, we might need to calculate the **expected return on a portfolio**. We do this simply by summing the weighted average expected return of the individual assets. The formula for this is:

(R) of a portfolio $= w_1 R_1 + w_2 R_2 + ... + w_n R_n$

where w_n is the weighting of a stock within the portfolio. So if 40% of our portfolio is made up of stock in Happy Farms with an expected return of 12%, 25% in Raaaah! with an expected return of 6%, and 35% in Motormouth Inc. with an expected return of 8%, the expected return on the portfolio is:

$$(0.4 \times 12\%) + (0.25 \times 6\%) + (0.35 \times 8\%)$$
$$= 4.8\% + 1.5\% + 2.8\% = 9.1\%$$

Just as with betting, we use variance or standard deviation to calculate how risky an investment is. To calculate variance (σ^2) for an individual investment, we look at each possible outcome and find the sum of the squares of the difference between the expected return for that outcome and the expected return, weighted by probability. The formula for this is

$$\sigma^2 = P_1(R_1 - E(R))^2 + P_2(R_2 - E(R))^2 + \ldots + P_n(R_n - E(R))^2$$

where P_n is the probability of the nth outcome, R_n is the return for that outcome and $E(R)$ is the expected return. Let's use the figures from Raaaah! above:

$$[0.2 \times (16\% - 6\%)^2] + [0.6 \times (6\% - 6\%)^2]$$
$$+ [0.2 \times (-4\% - 6\%)^2]$$
$$= (0.2 \times 100) + (0.6 \times 0) +$$
$$(0.2 \times 100) = 40$$

So the **variance** is 40, and **the standard deviation** (σ) is the square root of 40 (which is approximately 6.32). We can use the 68/95/99.7 rule (see p. 34) to get a sense of how often the actual return will fall in a particular range.

Beyond this, the formulae start to get a bit more complicated, but I will give a quick overview. For instance, to find the **covariance between two assets** (a and b) over a given period you first calculate the actual return from each day (or week or whatever) to the next for each asset. Then, for each of these periods, subtract the average return from the actual return for asset a, and multiply by the same figure for asset b – the covariance is the sum of these calculations divided by the number of periods, n:

$$\sum \frac{(R_a - \text{Avg } R_a)(R_b - \text{Avg } R_b)}{n}$$

Note that, in this formulation, a total period of x days' measurements contains $(x - 1)$ periods. Covariance is a measure of how movements of the two assets are related – a positive covariance means that the two assets tend to move in the same direction, a negative covariance the opposite.

Imagine Raaaah! has an average return of 6% and Motormouth Inc. has an average return of 8%. And we have the following actual results (I'm using a very small sample for the sake of simplicity).

Period	R_a	R_b	R_b - Avg R_b	R_b - Avg R_b	$(R_a$ - Avg $R_a)$ x $(R_b$ -Avg $R_b)$
1	5	7	−1	−1	1
2	12	1	6	−7	−42
3	8	16	2	8	16
4	3	9	−3	1	−3

Table 4. Covariance calculations for Raah! and Motormouth Inc.

So to calculate covariance we find the sum of the last column:

$$(-1 \times -1) + (6 \times -7) + (2 \times 8) + (-3 \times 1)$$
$$= 1 - 42 + 16 - 3 = -28$$

Then we divide by the number of periods, 4, to get the covariance of −7.

The **correlation coefficient** can also be used to measure the relation of the two assets. To find the correlation coefficient for Raaaah! and Motormouth Inc., we would divide the covariance of the two assets by the product of the standard deviation for each asset. The formula for this is:

$$\text{correlation coefficient} = \frac{\text{covariance}_a}{\sigma_a \sigma_b}$$

The correlation coefficient will be between 1 and −1. If it is 1 the two assets move in exactly the same way together, if it is −1 they move in exactly the same way but in opposite directions, and if it is zero they are totally uncorrelated. So, if you plot the price moves of two companies x and y on a standard graph, the single-day moves will tend to slope upwards from left to right (meaning that x and y are moving in the same direction) or downwards from right to left (meaning that x and y are moving in opposite directions).

The correlation coefficient is also useful when it comes to calculating the **standard deviation for an entire**

portfolio. This is a rather complex calculation. I've given the equation for a two-asset portfolio in a footnote,* for any readers who need to make this calculation, but will refrain from giving a long, boring explanation. The key thing to understand is that, given a good range of assets which are not 100% correlated, the standard deviation of the portfolio will be lower than the weighted average of the standard deviations of the individual investments – and how much lower depends on how correlated the assets are.

So while the mathematics does get a bit fiddly, it has the very useful purpose of establishing just how much risk there is in your portfolio – the greater the difference between the portfolio standard deviation and the weighted average, the more successfully diversified the portfolio is.

---

* If you are calculating the standard deviation for a two-asset portfolio you can use the following formula:

$$\sigma_{\text{portfolio}} = \sqrt{\left(w_1^2 \sigma_1^2 + w_2^2 \sigma_2^2 + 2 w_1 w_2 \rho_{1,2} \sigma_1 \sigma_2 \right)}$$

where:

w_1 = the proportion of the portfolio invested in Asset 1
w_2 = is the proportion of the portfolio invested in Asset 2
σ_1 = is the Asset 1 standard deviation of returns
σ_2 = is the Asset 2 standard deviation of returns
$\rho_{1,2}$ = is the correlation coefficient between the returns of Asset 1 and Asset 2

What to Do

Don't get too stressed if all these equations are making your head spin – there's a fairly limited number of occasions when you will need to calculate these figures. It's important to have a sense of the nuts and bolts of what the figures mean but there are plenty of online calculators or pieces of software that can be used to do the job for you. The important thing is to understand how a familiarity with portfolio theory can be useful in reducing risk.

Risk and Volatility

Just as with gambling, the volatility of an investment opportunity is directly related to how much risk and potential reward there is in taking it. The most common measure of volatility in the stock market is **beta**. There are various definitions of beta and more or less complex ways of working it out. But at heart it is a fairly simple idea. It is essentially a comparison of the volatility of a particular stock with the volatility of the market as a whole. This gives us a sense of how much of the variation in a stock's price movements is associated with general market conditions and how much is specific to the particular stock.

Covariance

An important concept that underpins the calculation of beta is covariance. Here's an easy way to visualize what covariance actually means in this context. Given a set of information for the prices of x and y on consecutive days, first create a scatter plot of these points. Then, rather than looking for a line of best fit for the plotted points, draw an arrow leading from the dot that represents the x and y co-ordinates for day 1 to the dot that represents the x and y co-ordinates for day 2, then continue to do this for day 2 and day 3, and so on. If you imagine this as a map, then when an arrow goes between 'southwest' and 'northeast' (regardless of whether it points downwards to the left or upwards to the right), the two sets of data have positive covariance, meaning that they will tend to respond to events by moving in the same direction; when the arrow goes between 'southeast' and 'northwest' then the two sets of data have negative covariance. When considering how to create a balanced portfolio that spreads risk, it can be useful to look at the covariance of the various elements of the portfolio. Where they mostly show positive covariance, it is more likely that the price of all of them will move in the same direction on any given day. If you want to use Excel to calculate variance, use = VAR.S (all the

percentage changes of the asset). For covariance, use
= COVARIANCE.S (all the percentage changes of the
asset, all the percentage changes of the benchmark).

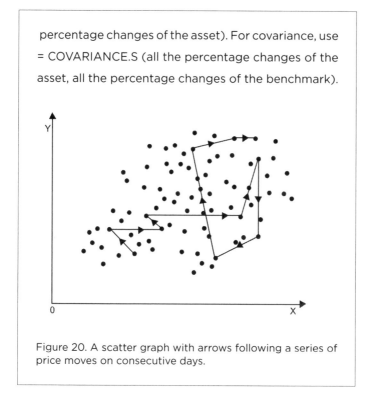

Figure 20. A scatter graph with arrows following a series of
price moves on consecutive days.

There are plenty of sources that publish values for the
beta of various stocks, but it is best to have a sense of how
the calculation is done, as published sources vary in their
methodology and the time period taken into account.

The easiest way to calculate beta is to compile a chart
or spreadsheet of the historical prices for a stock and a
benchmark market as a whole over a set period. Calculate
the percentage change from each period to the next
(whether daily, weekly, monthly or whatever) for both
the stock and the benchmark market. Then divide the

covariance of the market price changes and the stock by the variance of the price changes of the benchmark market.

The equation for this is:

$$\beta = \frac{\text{Cov}(r_a, r_b)}{\text{Var}(r_b)}$$

where β is beta, r_a is the return of the asset and r_b is the return of the benchmark market.

A beta of more than 1 roughly indicates that the stock is more volatile than the market. A beta of less than 1 indicates that the stock is less volatile than the market, while a negative beta value would imply (in a market with positive growth) that the stock lost value over the measured period.

Let's work through a quick example of that. Table 5 shows the returns from Motormouth Inc. and the market over a five-day period (expressed as a percentage of the previous day's closing price):

Day	Motormouth returns (%)	Market returns (%)
1	1.3	3.3
2	1.7	4.2
3	2.5	4.9
4	1.35	4.1
5	0.6	2.5

Table 5. Returns from Motormouth and the market

First, calculate the average return (mean) for each:

For Motormouth it is (1.3 + 1.7 + 2.5 + 1.35 + 0.7) / 5 = 1.49
For the market it is (3.3 + 4.2 + 4.9 + 4.1 + 2.5) / 5 = 3.8

To work out the variance of the market, add up the squares of the difference between each day's return and the mean.

$(-0.5)^2 + (0.4)^2 + 1.1^2 + (0.3)^2 + (-1.3)^2 = 3.4$

Next we work out the covariance. We take the differences between Motormouth's return and its average return, and multiply these by the difference between the market's return and average return. Then we take the sum of these calculations for the five days and divide the result by the sample size minus 1. The equation is:

$$\text{covariance} = \sum_{1}^{5} \frac{(RMo - ARMo) \times (RMa - ARMa)}{S - 1}$$

where
RMo is Motormouth's return
ARMo is the average Motormouth return
RMa is the market's return
ARMa is the average market return
S is the sample size

Substituting the values in the equation, we get:

$$\{[(1.3 - 1.49) \times (3.3 - 3.8)] + [(1.7 - 1.49) \times (4.2 - 3.8)] +$$
$$[(2.5 - 1.49) \times (4.9 - 3.8)] + [(1.35 - 1.49) \times (4.1 - 3.8)] +$$
$$[(0.6 - 1.49) \times (2.5 - 3.8)]\} / (5 - 1)$$
$$= 2.405 / (5 - 1)$$
$$= 0.60125$$

So, in this case, Motormouth has a fairly small positive covariance with the market – its price moves over this period were correlated with the market, but not to a very strong degree.

A more reliable method is the one used by the **capital asset pricing model**. Here the actual rate of the return of the stock and the benchmark market are discounted by the risk-free interest rate (usually the government bond rate.)

This leads to this slightly different way of calculating beta:

$$\beta = \frac{r_a - r_f}{r_b - r_f}$$

where r_f is the risk-free market rate.

This can be rearranged to find the predicted rate of return from a stock given its beta and the market rate of return.

$$r_a = r_f + \beta(r_b - r_f)$$

which essentially means that for a stock with beta higher than 1 we would need to see a rate of return higher than the market rate of return for it to be a sensible bet.

Whether you trust published beta values or prefer to use your own methods, benchmarks and timescales, the important thing is to take into account the level of risk as well as the time value of money when investing in a stock. As we saw in the gambling section, the greater the volatility of a bet, the higher the chances are of both a significant win and a significant loss (including the chances of going bust due to a freak occurrence).

The Hedge Ratio

When people use financial and investment jargon, it often turns out to be a needlessly complicated way to express a simple idea. The hedge ratio is a good example. It measures the **delta** of an option when you are using the option to hedge a position, which I'll explain using an example. Imagine that you run a road haulage company and want to hedge against future changes in the price of petrol. If you are able to buy **futures** (legal agreements to buy or sell something at a predetermined price at a specified time in the future) in the petrol, you can hedge as much as you want by simply buying them. Depending on the price by which the futures are discounted, this will give you something close to a perfect hedging position. However, it may be that you need to use a substitute such as futures in crude oil. Delta simply measures how much the price for crude oil will change for each unit of change in the petrol price. For instance if a $100 change in petrol prices will be accompanied by a $50 change in crude oil prices, this is a delta of 0.5.

If you are hedging using products which do not perfectly track each other's price movements, then it is best not to hedge the entire position. Instead you can use the optimal hedge ratio to minimize the volatility in your position. This is found by multiplying the correlation coefficient of the two positions by the ratio between the standard deviation of the spot price (the current price for which the product can actually be bought) and the standard deviation of the futures price. The equation for this is $h = p \times \left(\frac{\sigma_s}{\sigma_f}\right)$ where h is the optimal hedge ratio, p is the correlation coefficient and σ_s and σ_f are the standard deviations for the spot price and the futures price.

For instance, for the petrol and oil example above, if the standard deviation of the spot price for petrol is 4% and the standard deviation of the oil price is 8%, while the correlation coefficient between them is 0.9, you would ideally hedge $(0.9 \times \frac{4}{8}) = 45\%$ of your position. Note that the more closely correlated the two positions are, and the closer their standard deviations are to one another, the closer the optimal hedge ratio will be to 1, which would be a perfect hedge (meaning you should hedge the entire position).

This makes intuitive sense – if you can find a closely correlated asset to the one you want to hedge, it will make for a safer bet (and have less potential downside) than an asset whose price might not track your asset closely. Bear in mind also that hedging will reduce your potential profits as well as limiting your potential losses. If for instance you hedged your projected petrol requirements with crude oil futures and the price of both oil and petrol fell, then you would have lost out on potential profits. So hedging is essentially about risk reduction.

I've mentioned that hedge funds were originally named for their use of this kind of investment tactic, combining long and short positions to reduce risk – but they always used a combination of other strategies too, and the methods they employ to try to find an edge in the market have expanded considerably over time.

Arbitrage

Arbitrage is a common strategy in business as well as gambling. The equivalent strategy in business comes when an investor tries to exploit small differences in pricing across the markets.

In a sense, most businesses that buy and sell products can be seen as using a form of arbitrage – a shopkeeper who bulk-buys his grocery products in the cash-and-carry warehouse and then sells them in smaller quantities to customers is exploiting the price differential and the fact that his customers are willing to pay for the convenience of having these products for sale locally.

However, arbitrage technically just refers to the simultaneous buying and selling of assets in different markets (or in derivative forms) to take advantage of differing prices for the same asset. Just as 'arbers' can sometimes find bookies offering different odds on the same event, financial arbers are looking for temporary inefficiencies in the market which allow them to buy and sell at a guaranteed profit.

If for instance you happen to notice that stock in Motormouth Inc. is trading at $38.47 in the Tokyo stock exchange, but at $38.52 in London, you can in theory buy stock in Tokyo and sell it instantaneously in London as long as the price differential remains. There are a significant number of businesses and sole traders around who also look for arbitrage opportunities in specific products – for instance buying a copy of a book on Amazon and instantly selling it on for a higher price on eBay.

Another more complex example comes from the currency markets. Imagine a situation where you can get the following rates for euros, pounds and dollars in different exchanges around the world:

Exchange 1: 1.45 dollars to the pound
Exchange 2: 0.8 euros to the dollar
Exchange 3: 0.88 pounds to the euro

If you start out with £100, you could first convert this to $145, then to €116 and finally back to approximately £102.08, making 2% profit on the trade. £2.08 won't go far, but if you had started with £10,000 instead, the profit would be more significant at £208.

Obviously such inefficiencies across markets are not easy to find, and will quickly be corrected as funds around the world buy and sell the currencies or other assets that are out of line, pushing the prices back into equilibrium. And you need mechanisms to ensure the trades really are simultaneous, and to make enough profit to pay for any commission. However, the basic principle of arbitrage underlies many of the activities of hedge funds and other speculators. In particular, automated trading exploits tiny variations in price to generate a large amount of small but profitable trades.

Bulls and Bears and the Rationality of Markets

There are many examples from history of speculative booms and crashes. One early example was the tulip mania in seventeenth-century Netherlands in which the price of tulip bulbs escalated to extraordinary highs before collapsing (the story has been somewhat exaggerated in some historical accounts, but it was nonetheless a real phenomenon). The Panic of 1873 was a financial crisis that triggered the Long Depression in the United States and Europe; the Wall Street Crash similarly led to the Great Depression, while the boom in property derivatives was largely to blame for the global financial crisis of recent years. Along the way there have been hundreds of local booms and busts in markets, from property to stocks to derivatives to goods.

Throughout the market cycle you can find bulls, people predicting that prices will rise, and bears who predict the opposite. It can be a passionate debate as it is never easy to know where we are in the business cycle. From a mathematical point of view it's interesting to consider some of the underlying driving forces that make it hard to say what is rational.

The idea of a Keynesian beauty contest comes from the writings of John Maynard Keynes. He discussed a situation in which contestants in a game were asked to pick the six most attractive faces from 100 photographs. The winner will be the one whose guess most accurately predicts the average answer given by entrants. Keynes points out that rather than choose the six faces you personally find most attractive, the rational behaviour in this case would be to pick the six that you believe will be found most attractive **on average**. And it gets even more complicated... Since this is the rational behaviour, you have to assume that everyone else will also behave that way. So you are actually trying to guess which six faces the average entrant will think the average entrant will find most attractive... So there is a kind of feedback loop in how we define the rational behaviour.

Keynes' point was that the stock market behaves in a similar way. The most successful stock pickers will often be the ones who manage to predict what the market as a whole thinks about particular stocks, rather than attempting to find their true value. So while anyone can use equations such as the price to equity ratio or the Gordon equation to try to analyse companies, part of the thought process also has to be to work out what other people will make of the same publicly available information (especially the largest funds, whose decisions will have the most impact on the market). And then you need to work out what they will think other people will make of that information, and so on.

To understand how mathematically complex such a feedback loop can be, we need the concept of a **Nash equilibrium**. This is a part of game theory, a branch of maths that tries to analyse all the possible moves in a competitive game involving two or more players where each player knows the equilibrium strategies of the other players.

Let's investigate one such game. The so-called *p*-**beauty contest** is a variation on Keynes' version. Entrants to the game are asked to guess which number from 0 to 100 will be $\frac{2}{3}$ of the average of all the guesses that entrants to the game have given.

Clearly there is no point guessing any number higher than 66. Now, if we were to start by assuming an even distribution of random guesses between 0 and 66, we'd expect a mean of 33, so it might be reasonable to start by guessing 22 which is $\frac{2}{3}$ of 33. And we can assume that others will come to the same conclusion. However, if we reason this way and expect all other players to guess 22, we would be pushed towards guessing 14, which is roughly two thirds of 22. And if we expect all other players to spot that logical step and to guess 14, then it would make $\frac{2}{3}$ of 14. And so on. It turns out that the only Nash equilibrium solutions for this problem when it is taken to a logical conclusion are for all players to guess 0 or 1. However, in practice some entrants are not perfectly rational (or will assume that the other players are not all perfectly rational) and will not follow the thought process to this extreme conclusion, so while this is a Nash equilibrium for the problem it may not be the actual result in a real-world experiment.

This is borne out by the times people have performed this as a psychology experiment. Players can be defined by how many steps they follow through the potentially infinite feedback loop – a level 0 player will guess in the range of 33, a level 1 player will guess 22, and so on. When the experiment has been performed, most players range from level 0 to about level 3 and the average guess comes in somewhere between 15 and 20 as a result.

In game theory literature, these levels are sometimes referred to as a measure of the participants' 'depth of reasoning'. But this seems incorrect because the smartest guess is one which takes this likely layering of levels into account – the winner of the contest is the one who guesses what level of reasoning will be average, not the one who assumes everyone will see the potentially infinite feedback loop.

In some ways, business in general resembles the Keynesian beauty contest. While it would be nice to think that book publishers just publish the books **they** believe to be great, it is only rational for them to publish books that they believe a lot of bookbuyers will buy (which in some cases may mean trying to appeal to what they think of as the lowest common denominator). And, beyond that, they need to believe that their sales force will believe that book buyers will like this book. And the sales force need to believe that the retailers will believe that their customers will like this book. So there is already a feedback loop of several layers going on in this market.

Going back to bulls and bears, it's easy to think that markets are irrational. When there is a boom, people act as though prices of a particular asset can rise at a rate higher than inflation forever. When the price inevitably stalls or crashes, fear seems to kick in, and prices fall to what seem like irrational lows. However, if you bear in mind the concept of the Keynesian beauty contest, it is equally reasonable to see that as the inevitable result of a marketplace in which everyone is trying to guess exactly what other people might be thinking. And the combination of a lot of rational guesswork can still be a market that does crazy things, such as pricing tulip bulbs at the same level as a house in the period of the Tulip mania.

Incidentally, it's worth relating this thought to the curious fact that many investors and financiers claim to believe in the efficient markets hypothesis. This idea suggests that, because everyone can use the same information to value stocks, these always trade at their fair value (because as soon as a disparity exists between the fair value and the market value someone will recognize this and buy or sell the stock until it returns to the fair value). This in turn must imply that it is more or less impossible to buy an undervalued asset. That's hard to square with the fact that many of these same people believe they can personally beat the market.

Economists such as Robert Shiller, the author of *Irrational Exuberance*, have argued, to the contrary, that the market is too volatile to be explained by the efficient markets hypothesis. In the end the simplest summary of the situation might be that, in light of the logic of the Keynesian beauty contest, it's not truly possible to define the correct value of an asset, so people behaving in a perfectly rational manner can still combine together to form a highly irrational marketplace.

Kelly Betting for Investors

Some of the world's most successful investors, including Warren Buffett and Bill Gross, have relied to some extent on the Kelly formula (see Chapter 3) in building up their fortunes. A good understanding of the system would be useful to investors as well as gamblers. Of course this is far from being the only tool these investors use, but it is extremely useful when it comes to deciding how much of your capital to stake on particular assets. Warren Buffett's partner Charlie Munger has talked about the importance of being careful how thinly you spread your investment capital: 'the wise ones bet heavily when the world offers them that opportunity. They bet big when they have the odds. And the rest of the time, they don't. It's just that simple.' When it comes to investing, we can use a simplified version of the criterion, to deal with the fact that a losing investment generally won't be a total write-off (as bets tend to be). The equation in this case is

$$\text{Percentage of capital to invest} = \frac{\text{total expected return}}{\text{fractional odds}}$$

To calculate the total expected return on an investment, find the probability of each outcome, and multiply by the percentage gain or loss in that outcome, then add these together. For instance, you might want to look at 100 recent trades in a similar asset and make a table of the outcomes:

Number of occurrences	Gain
25	6%
20	2%
40	0%
15	-4%

Table 6. Outcomes of 100 recent trades

The expected return from our next similar trade would be:

$$(0.25 \times 0.06) + (0.2 \times 0.02) + (0.4 \times 0) - (0.15 \times 0.04)$$
$$= 0.013$$

So we have an expected return of 0.013 = 1.3%.

To calculate the fractional odds, compare the number of losing to winning outcomes. We have 15 losing outcomes and 45 winning ones, so in this case the odds can be expressed as 1 to 3 or $\frac{1}{3}$, so we divide the total expected return by 0.33: $\frac{1.3\%}{0.33}$ = 3.9%.

So, for this scenario, it is worth investing up to 3.9% of your capital in the asset. (Bear in mind that 'short odds' represent a likely outcome, whereas 'long odds' represent an unlikely one. So if a particular asset lost money in 75% of trades and won in 25%, we would divide by 3 and come to a smaller suggested investment, which makes intuitive sense. If you find it more intuitive, you can use the decimal odds minus 1.)

Of course, as with gambling, this can only provide a very rough rule of thumb for investors, but it is a useful tool to work with as it provides discipline over how much of your capital you invest, according to how risky the investment is.

There is some fairly academic dispute about whether this criterion really works for investment, and Warren Buffett may simply be the luckiest guy out there who keeps winning. But I tend to go on the theory that if it's good enough for Warren Buffett, it's good enough for me.

Low-Risk, High-Potential Investing

Investing in stocks or derivatives isn't for everyone. But if you do attempt to make your money this way, it's important to have a grasp of the basic mathematical concepts involved. We've seen a few ways to measure the value of a company, to spot assets that have high potential for growth, and to balance a portfolio in order to minimize risk. Most successful investors use a range of such methods and come to understand them thoroughly over a period of time. So while it is worth finding out what measures have worked for millionaires such as Warren Buffett, there is no substitute for getting the experience of using financial tools on a daily basis.

Buffett, often referred to as the 'Sage of Omaha', claims his approach to value investing is '85% Benjamin Graham'. Graham was an economist, professional investor and author of the classic books *Security Analysis* and *The Intelligent Investor.* He essentially invented the idea of value investing, having taught it at Columbia Business School in the 1920s.

He used to tell a parable in which an individual called Mr Market knocks on your door each day offering to sell you shares in companies. Most days the price he quotes seems fairly sensible, but on a few days the price seems far too high, while on a few others it seems too low. As with the Keynesian beauty contest, it is not necessarily for us to say whether the price movements are rational or not, but the swings in price will often seem excessive.

The point of the parable is to warn against taking the vagaries of the markets too seriously and always to focus on the concept of intrinsic value when investing. Just as a gambler looks for situations in which the house edge is reversed and they can bet at an advantage, the smart investor looks for situations in which the market price for a share is lower than its intrinsic value (at which point it is sensible to buy) or vice versa (in which case it may be time to sell up).

One of Warren Buffett's range of formulae for analysing intrinsic value was originally suggested by Graham. Looking at how sensibly a company invests in its own future success, he compared the retained earnings (in other words earnings that weren't paid out as dividends) over a five-year period to the increase in value in the total shares. The basic rule of thumb is that the company is investing retained earnings wisely if the increase in share value equals or exceeds the total retained earnings. Buffett gives a simple reason why this is a common-sense approach:

> You should wish your earnings to be re-invested [by the company] if they can be expected to earn high returns, and you should wish them paid to you if low returns are the likely outcome of re-investment.

In other words, if the company can do better with your money than you can yourself, it is reasonable for them to retain and reinvest the profits, otherwise they should hand them over.

Buffett has recently acknowledged that this rule of thumb doesn't work in periods such as 1971–5 and 2009 when the stock market falls sharply and there are less retained earnings to be measured, in which case the comparison has to be with the percentage by which the company's share value trajectory has exceeded the market average, rather than the simple increase in share value. And it might be smart to do this even in rising markets, so that a company in a bull market doesn't appear to be investing wisely simply because its results have kept pace with rising stock prices.

As with the other investment formulae we have mentioned, this measure should only be considered alongside other measures of a company's prospects when considering investing. However, if you are considering letting a company look after your money, it is fundamental to ask whether the money retained by the management of the company is being used as well as, or better than, you could do yourself.

What Not to Do

We can generalize from the Sage of Omaha's rule of thumb to a really fundamental mathematical rule of managing your money, one that is so simple that it almost doesn't need stating: **Never let someone else look after your money, unless the rate of growth they can achieve (after they have charged any fees or deductions) is higher than the rate of growth you could achieve yourself.**

The Irrational Investor

In the previous chapter we discussed some of the biases and pitfalls that can lead gamblers to make irrational decisions. However, it's not just gamblers who are subject to delusions and confusions. Investors and businesspeople are just as prone to falling prey to cognitive biases, including the optimism bias, the availability bias, the hot-hand fallacy and a host of other errors. From the boss who promotes a salesperson based on one year's lucky figures, and the fund managers who underrate the role of luck in their results, to the investor who thinks they are 'due' a win following a run of losses, there isn't a single logical bias that doesn't apply as much to business as it does to betting.

Author Nassim Nicholas Taleb has written about the 'black swan', an event that seems incredibly unlikely simply because it has never happened before. His conclusion is that people rely too much on inductive reasoning based on past events when it comes to predicting the future. The collapse in the sub-prime market which led to the global financial crisis was treated by analysts in advance as a 6-sigma event, meaning that it was more than six standard deviations from the mean and thus had a probability of about 2 in a billion. This is an example of how excessive faith in mathematical modelling can lead to a disastrous outcome.

It's also worth mentioning that there are some investment formulae and methods around which have a pseudo-mathematical basis but are not much better than voodoo or the most dubious of gambling 'systems'. A particular culprit is **chartism**, in which investors superstitiously believe they can understand and recognize patterns in stock charts. Chartists claim to be able to predict the future by observing shapes such as the 'head and shoulders' or the 'cup and handle' in these charts. It's a fairly obvious example of how humans make the mistake of looking for patterns in random information, and most chartists are pretty bad investors unless they have some much more useful means of analysis against which to check their hunches.

Dow theory is another example of spurious reasoning couched in technical-sounding terms. With its reference to 'up trends', 'down trends' and 'corrections', it has the vice of giving investors a false idea that the future can easily be predicted from the past. This leads to dodgy ideas such as the **fifty percent principle**, which is the idea that upward trends will be interrupted by falls of about 50% of the gains – this is taken to be normal behaviour, which will normally be followed by further gains, while losses above this figure are a sign of worse to come. This can lead to overconfidence in investors who rely on the theory, as it encourages them to look for patterns that may not be there.

What to Do

Do use genuine mathematics and use other common-sense tools in analysing investment opportunities. But avoid systems and methods that claim to be able to find patterns where there may be nothing but randomness. And even when you are using more serious methods of analysis, always beware of drawing large conclusions from a small amount of data.

Chapter 4 Summary

1. If you want to make money from the stock market, then make sure you understand the absolute basics of how various assets work.

2. Use financial ratios and other mathematical methods of analysis where you have a strong grasp of their operation.

3. Avoid nonsense such as chartism, and make sure you base any analysis on a large amount of data, since it is easy to perceive patterns in randomness if you use insufficient information.

4. Portfolio theory may be hard work (and, frankly, a bit dull), but it can help you to effectively reduce your risk and to make sure that your diversification is as effective as possible.

5. If you know what you are doing, hedging, arbitrage and the Kelly criterion can all be effective moneymaking strategies.

6. When everyone in a market is trying to second guess what everyone else is thinking, the mathematical feedback loop created can lead to behaviour that seems highly irrational and is genuinely hard to predict.

CHAPTER 5

Hacking, Cracking and Gaming the System

The general who loses a battle makes but few
calculations beforehand. Thus do many calculations
lead to victory, and few calculations to defeat...
It is by attention to this point that I can foresee
who is likely to win or lose.

Sun Tzu

Up to this point we have only discussed how mathematics applies to straightforward gambling, business and investment practices. But there are plenty of examples of people who have found ways to use maths to try to get a more dubious advantage. Some of these are obviously more legal or justifiable than others. However, let's stick to discussing the mechanics of these methods and systems rather than moralizing about them. For my own purposes, I think a reasonable question to ask when it comes to the morality of a particular hack or crack might be this...

What Would Ed Thorp Do?

As a mathematics professor, gambler and founder of the world's first quant hedge fund, Ed Thorp is a fascinating and inspiring figure. (Quant is short for quantitative analysis, and refers to the use of computers for financial analysis.) While he was working towards his PhD, Thorp became interested in the Kelly criterion and learnt how to programme early computers in order to experiment with strategies that might help a gambler. Having proved to his own satisfaction that the house edge for blackjack could be beaten by cardcounting, he approached the wealthy gambler Manny Kimmel and, using $10,000 of Kimmel's money, headed off to the casinos of Las Vegas, Reno and Lake Tahoe to test the theory. He succeeded in more than doubling his money in the first weekend. He continued to test out his theories (which also extended to other games such as baccarat and backgammon). He was often banned from casinos and had to resort to disguising himself with fake beards and dark glasses, and even claims to have been drugged on one occasion by staff of a shady casino. In 1962 he published *Beat the Dealer*, the bestselling book in which he made his cardcounting methods known to the wider public. The casinos were consequently forced to introduce extensive measures to counter the practice.

Thorp also carried out a wonderfully eccentric experiment with Claude Shannon (a successful mathematician in his own right, who was well known for his codebreaking during the war). They designed the world's first portable computer, which Thorp wore strapped to his leg, with fragile wires running to an earpiece connecting him to Shannon. They came up with a way of predicting where on a roulette wheel the ball was likely to fall, from the speed and trajectory with which it was cast – it was not a perfect prediction, but they could narrow the section of the wheel down to about six of the thirty-eight numbers, thus gaining a significant edge. In the space of an hour in experimental conditions, betting $25 per number, they won a notional $8,000. The computer worked well in laboratory conditions, but when they attempted to use it in a casino, there were operational problems (such as the audio wire coming loose in the ears of the player or being too visible) which prevented them from putting it to more extensive use. In any case, this was a mere experiment to Thorp, an academic exercise rather than a pure moneymaking scheme.*

Over time Thorp became more interested in the financial markets, realizing that the same methods he had pioneered – using computers to analyse patterns of probability – could be applied to a wider sphere. He concentrated on the analysis of small correlations and anomalies in pricing which could be exploited for reliable profits. He published

---

* In the 1970s a small group of graduate physics students from University of California Santa Cruz called the Eudaemons built an improved version of the roulette predictor – they were attempting to raise some money to fund a scientific community. Their experiment was somewhat successful, in that they won $10,000, although problems with the device again led to them finally abandoning the project.

Beat the Markets, a sequel to his earlier book, in 1967. And in 1969 he set up the Princeton Newport Partners fund, which is regarded as the first quant hedge fund. In the first eighteen years, he turned an initial $1.4 million fund into $273 million, making a profit in every single quarter. Of course this was largely the result of being a pioneer – now that there are so many quants and so many funds relying on computer analysis, it is much harder to find the kind of edge that he was initially able to identify in the markets.

Following the global financial crisis Thorp has been trenchant in his criticisms of Wall Street – he regards it as no better than a casino, arguing that the big banks are effectively able to make their own rules and distort the markets in their own favour. He also claims to have warned people about Bernie Madoff years before his final downfall, having recognized that many of Madoff's claimed trades were in fact impossible or hadn't happened. Madoff's status as a Wall Street insider combined with regulatory incompetence served to ensure Thorp's warnings were ignored.

Thorp is also notable for the variety of people he has inspired. Warren Buffett advised investors in his own hedge fund to join Thorp's instead. Bill Gross, the founder of the successful American investment management firm Pimco, originally set up as an investor after reading *Beat the Markets*. Ken Griffin, who set up the Citadel fund in 1990, originally relied on advice and documents from Princeton Newport.

It's well worth reading some of Thorp's books for inspiration. A couple of pieces of advice from him stand out. First, while explaining why he was successful at investment, he notes that people simply aren't 'good processors of information'. He says they are too easily distracted by statistical noise and 'fake news'. As a result, anyone relying on really strong, properly analysed information can have a genuine edge. He also notes that when it comes to more traditional investment, unless you are really going to do the legwork of understanding the companies you invest in (as Warren Buffett does, for instance) you should just stick to index funds. Because, according to his analysis, those who try to beat the market actually do 2% worse on average than the market itself. So, unless you have some genuine information to go on, you could make twice as much money over thirty-five years by indexing rather than relying on other systems and hunches.

Going back to the start of Thorp's career, let's take a look at the art of cardcounting next.

An Introduction to Cardcounting

One of the most common card games in casinos around the world is blackjack. And the really curious thing about blackjack is that it is a game that is eminently beatable by a sufficiently skilled cardcounter, and the casinos have known this, at the very least, since the publication of *Beat the Dealer*.

The first two points to make about cardcounting are that it is neither as difficult nor as easy as it might sound. The actual systems used to count cards do not involve any great mathematical skill or astounding feats of memory. Do not imagine that you need to be like the autistic savant in *Rain Man* who is able to memorize the exact order of six packs of cards: the process is far simpler than that. However, you do need great reserves of patience, calm, and an ability to stick to the system without getting rushes of blood, and you also need to know about the many ways in which casinos attempt to discourage or defeat cardcounters.

Cardcounters may well have been raking in profits from the early days of blackjack (it was only a legal game in the US from the 1930s, but had been around in one form or another in Europe and America since at least the sixteenth century). The so-called 'Four Horsemen of Blackjack' (Baldwin, Cantey, Maisel, McDermott) wrote a book in 1957 called *Playing Blackjack to Win* which discussed keeping track of cards as a way to tilt the game in your favour. But it was Thorp's *Beat the Dealer* that reached a much wider audience, making many more gamblers (and casinos) aware of the mathematics of cardcounting.

The basic principle behind cardcounting relies on the fact that when the remaining deck (or combined set of decks) of cards contains a high proportion of low cards (from two to seven) it benefits the dealer, while a deck containing a high proportion of cards worth 10 (ten to king) or aces (depending on the system) benefits the player. Thorp, who was looking at the traditional single-deck deal, analysed this mathematically and recommended a 'ten-count' system in which you start from a count of the sixteen cards worth 10, and of the thirty-six other cards, then divide the latter by the former to keep track of how the dealer's edge was varying.

Most subsequent cardcounters have concluded that it is better to use a simpler system, especially given the complications introduced by dealers working with six packs of cards instead of a single deck (a measure introduced specifically to make cardcounting more difficult). For instance, the basic 'Hi–Lo system' allots a value of 1 to cards two to six, a value of 0 to cards seven to nine, and a value of –1 to cards from ten to ace. All the cardcounter needs to do is to add up the value of the cards dealt and to keep a running count of the positive or negative total. (There are also other cardcounting systems you could choose, with names such as Omega, Hi-Opt and KO. I'll stick to Hi–Lo in this section as it is the most popular and has been shown to be relatively effective.)

How to interpret this total depends on other factors, for instance how many decks of cards are being used. There are very few casinos these days playing with a single deck of cards, as it makes cardcounting too effective. Instead they will play with six decks or similar. To get the 'true count' from your running count, the quickest method is to estimate the number of decks left and divide by that number. The intuition behind this is that the balance of high and low cards will be more significantly impacted when there are a smaller number of cards in play.

At its simplest, the strategy is to bet a set lower amount when the count is negative, and a higher amount when it is positive (note that the values for an entire pack add up to zero, so this is what is known as a balanced system, where a count of zero means a remaining pack that has no advantage either way). Or you might want a slightly more complex system like increasing your bet by 1 unit for each unit the count increases by.

Of course there are more precise and complex methods available which can improve your odds. However, bear in mind that casinos study the most popular books and methods and are aware of specific betting schemes that have been heavily used. So if you use one of these systems you are more likely to draw attention to what you are doing.

There are also more complex mathematical strategies that can be used. For instance, there has been some expert analysis of how the odds vary after the first few cards have been dealt, and which situations you should bet more or fold in. Two excellent books on such detailed strategies are *Blackjack Attack* by Don Schlesinger and *Professional Blackjack* by Stanford Wong. The latter author gave his name to Wong Halves, a complicated counting system in which the cards worth ten and the ace are counted as -1, cards two and seven count $\frac{1}{2}$, cards three, four, six count 1, card five counts $1\frac{1}{2}$, card eight counts as zero and card nine as $-\frac{1}{2}$. This has a high level of accuracy to the actual odds, but is obviously more difficult than the Hi–Lo system.

Whatever system you choose to learn, you need a lot of practice to be able to count cards effectively. You need to be able to apply the system at great speed in a casino where there are plenty of distractions and background noise to make it more difficult.

Also casinos go to a lot of trouble to spot and discourage cardcounting. If you try it, you may find the dealer and pit boss giving you suspicious glances, casino employees talking to you to throw your concentration, and CCTV of your play and your betting patterns being analysed. There has even been academic research into experimental software that would automatically spot cardcounters by analysing their hands and betting patterns. The casinos tend to employ ex-cardcounters who are good at spotting the signs. And you can end up banned not only from the casino you are playing in, but others as well, as casinos share footage and images of known counters.

What to Do

If you want to try cardcounting, learn a system, and practise it until it is second nature. Read up on the ways in which casinos try to combat cardcounters and prepare for these. And assume that the natural variance of luck will mean you need a large pot in order to use cardcounting strategies with any success.

Remember that, while anyone can learn the basic systems, some people have a better temperament for it than others, and the variable legality of the practice in different regulatory regimes means you may need to be willing to sail close to the wind.

Cardcounting can genuinely reverse the edge in a game so that the gambler has a slight edge on the casino. On its own this doesn't guarantee success and the methods used by casinos to stop cardcounting have made it harder than it used to be, but a skilled cardcounter can still come out on top.

The MIT Blackjack Team

One of the most flamboyant groups to use cardcounting in a methodical way was the MIT Blackjack team. During the 1970s, cardcounting had been banned at the casinos of Atlantic City, but towards the end of the decade a court ruling prevented the casinos from doing this (officially, at least). A group of six students from Massachusetts Institute of Technology (MIT) taught themselves to count cards and went on a highly lucrative spring-break trip to take advantage of their newfound skills. The original group disbanded, but one of them, J. P. Massar, would go on to more systematic moneymaking schemes.

Having taught cardcounting as a course at MIT, Massar helped form a group which was initially made up of four players. Then, in the 1980s, he had a fortuitous meeting with Bill Kaplan, a Harvard graduate who had set up his own blackjack team a few years earlier. Kaplan's team had had a highly profitable run at Las Vegas before their activities became known to the casinos, at which point they had been forced to break up into smaller teams and seek new casinos abroad at which to operate.

Kaplan agreed to help fund a new team but suggested a far more rigorous approach in which the strategies were all fully agreed and worked out in advance. He raised just under $200,000 from a range of investors, and the team started to operate in August 1980. In the initial period they were making the equivalent of a 250% annual return.

The MIT team would go on to recruit up to eighty players who would play in smaller units, and continued to make a very healthy return on their investments and start-up capital through the 1980s and into the following decade. By 1984 Kaplan couldn't be seen in a casino without being followed by casino staff trying to spot his teams. But by bringing in a constant supply of new faces they were able to partially evade the casinos' efforts at detection. The systems they used for betting were relatively simple, but improved on individual cardcounting by having an independent counter who was separate to the gamblers and could thus concentrate on that task alone. The counter would then use signals such as hand gestures when the situation was positive.

The key lesson that can be learnt from the MIT Blackjack team is that cardcounting can still be profitable if you are careful about how you operate and that an ingenious way of making money can often be taken to new heights by someone who applies it in a systematic and focused fashion.

Dice and Card Cheats

Since the dawn of gambling, the less scrupulous have been looking for ways to get an illegitimate advantage, meaning that games of dice and cards have frequently been subject to sharps and cheats. Some of these can be spotted or evaded using a bit of mathematics. For instance, the most basic way of cheating at dice is to swap the real dice for a 'top' (a dice with only three numbers on it, with each number repeated on opposite faces) or a 'one-way top' (where only one number is repeated). In the game of craps, such dice can be used to increase or decrease the odds on certain throws. For instance, using two 1–3–5 tops would make it impossible to throw a 7 and 'crap out'. It isn't easy to instantly spot a top as you only clearly see three faces of the dice at a time.

If you examine a standard dice, they have a regular pattern – meaning, for instance, that if you hold the dice with the faces with one, two and three dots showing, they will run in a counterclockwise direction, as below.

If you now take a blank cube and turn it into a top by adding 1, 2, 3 in this formation on three faces, and the same numbers on the opposite faces, you will find that in a number of orientations these three numbers will run in the 'wrong' direction. It's mathematically impossible to avoid this problem when you make a top.

Dice odds can be confusing. The great mathematician and philosopher Gottfried Leibniz made an easy mistake when he stated that the odds on throwing an 11 with two dice are the same as the odds of throwing a 12. His thinking was that each is only feasible with one combination of numbers, (6, 5) for 11 and (6, 6) for 12. However, you have to take into account the fact that you can throw a 5 with either dice and a 6 with either dice – so while there is only one way to throw a 12 (6, 6) there are two ways to throw an 11: (5, 6) and (6, 5).

As it is so easy to make this kind of error, a common trick for hustlers is to offer to take a bet on something which seems intuitively to be a good bet, but for which it is easy to miscalculate the odds.

For instance, one con involves making an even money bet that the mark (the victim of the con) will throw an 8 before a 7. Most dice players know that it's easier to throw 7 (which can be made with six combinations: 1, 6; 2, 5; 3, 4; 4, 3; 5, 2; 6, 1) than 8 (for which there are only five: 2, 6; 3, 5; 4, 4; 5, 3; 6, 2), so the odds are in the mark's favour. Assuming the mark wins the first small bet, a second even money bet is offered on the mark throwing a 6 before a 7. There are also only five ways to throw a 6 (1, 5; 2, 4; 3, 3; 4, 2; 5, 1), so again the odds are in the mark's favour. At this point the con comes into play – the mark is offered a much bigger even money bet if they can throw a 6 and an 8 before they throw two 7s.

This feels like an expanded version of the same bet, but actually the odds now shift in the hustler's favour – they are 4,255 out of 7,744. If the hustler had specified which order the 6 and 8 should fall, they would still be offering a losing bet: it is because the 6 and 8 can come in either order that the odds have shifted.

To see why, look at Figure 21, which illustrates what happens at the start of a series of two throws of two dice. If the first dice falls on the number in the first column below, the middle column shows which numbers will make up a 7, while the right column shows which will make up a 6 or an 8. The right column is significantly more probable at this stage: 10/36 as opposed to 6/36. Once the first

6 or 8 has fallen, the odds on the second of the two falling go back to 5/36, but the advantage from the initial phase of the series of throws is enough to shift the odds in the hustler's favour overall.

Dice 1	Dice 2 (for a 7)	Dice 2 (for a 6 or 8)
⚀	⚅	⚄
⚁	⚄	⚃, ⚅
⚂	⚃	⚂, ⚄
⚃	⚂	⚁, ⚃
⚄	⚁	⚀, ⚂
⚅	⚀	⚁

Figure 21. How to throw a 6, 7 or 8 with two dice.

There are some similar propositions that use cards rather than dice. Maverick solitaire is a game named after the US television show *Maverick*, in which the gambler Bart Maverick takes a huge bet which he will win if he manages to arrange twenty-five cards, picked at random from a deck, and arrange them into five sets of five, each of which is a straight or higher at poker. (So each must be a straight, flush, full house, four of a kind, straight flush or royal flush.*)

---

* If you need reminding: a straight is a run of cards in numerical order (with jack, queen, king counting as 11, 12, 13), a flush is a set of cards from the same suit, a full house is three of a kind plus a pair in the same hand, a straight flush is a straight with all the cards from the same suit, and a royal flush is 10, jack, queen, king, ace from the same suit.

This intuitively seems like it would be hard, but if you try it out you'll find it is usually fairly easy. For a start, the least number of flushes you can get is a selection with four cards from two suits and nine from the other two, but even in this combination you have two guaranteed flushes, and a lot of flexibility in which five cards you use from those two suits. Then you can move cards in and out of the flushes, and thus start putting together some straights and full houses pretty quickly. The only situation, for instance, in which you don't have at least one full house would be if you had two of each value except for one, for which you have a single card. But in this scenario, you would have a big choice of straights to choose from.

In fact, if you did the complex combinatorial maths, using Pascal's triangle (or, preferably, a computer) you'd find that the odds on being able to achieve this feat are just over 98% – so if you want a real challenge it's instead worth trying to find the sets of 25 cards that **don't** allow you to succeed at Maverick solitaire.

Another card-based proposition bet is to take two decks of cards and shuffle each thoroughly. Offer a friend (or mark) the following even-money bet – you will each turn over one card at a time, and if the two cards showing are exactly the same rank and suit you win, whereas if you get to the end of the pack with no match they win. The odds of each pair of individual cards matching are only 1/52 so your friend will tend to see this as a good bet. In fact you have an edge of approximately 26% in this game.

A similar gamble is the royal bet – the hustler offers you even money on the following proposition – you cut a deck of cards into three piles, then the hustler bets that at least one of the cards you draw from each will be a jack, king or queen. Given that there are only 12 of these in a pack of 52, the odds seem to be in their favour on a casual inspection. In fact the odds here are cumulative – you have a 12/52 chance on the first cut. If that isn't a royal you have a 12/51 chance on the second cut, and then 12/50 on the third. So you have 12/52 + 12/51 + 12/50 = 70.6% chance of finding a royal and you will only fail to find one on about 30% of the draws.

You can take two morals from these kinds of hustles. One is that if someone offers you a bet that sounds too good to be true, it probably is, so think very carefully about the actual probabilities involved. And the other is that a bit of mathematical nous can sometimes allow you to get an edge with no actual cheating, just a basic grasp of probability.

Probability Scams

One of the most basic scams that relies on mathematics is the baby predictor. A relatively cheap service is offered whereby the scammer claims (through psychic ability or other means) to be able to predict the sex of your baby, with the guarantee of a refund if they get it wrong.

Of course all the scammer need do is make guesses at random, keep the money from the 50% of their predictions that will be correct, and refund those of the remaining 50% who remember to ask for it.

The same basic principle is behind the well-known 'perfect prediction scam'. The scammer picks an event that has a binary outcome, such as a knockout soccer match, or whether the stock market will rise or fall in a given week. (Alternatively, a similar scam can work for events with very limited outcomes, such as three-horse races.) They then send out 16,000 letters or emails claiming that the scammer has a perfect prediction method or inside information, half of which predict winner A and half winner B. The following week they follow up on the 8,000 that had the winner with a new prediction, again, split half and half. After four rounds of this they end up with 1,000 people who have received four correct predictions in a row. Now the scammer offers to continue sending predictions to these people, in return for a fee of, say $100. If even 10% of the 1,000 accept the offer, the scammer has made $10,000 (minus the costs of setting up the scam). In the TV documentary *The System*, the magician used a version of this scam, 'predicting' five horse-racing results in a row before persuading one woman to part with £4,000 to bet on the sixth race – at which point he revealed what he had done. (There was a further twist: it was a losing bet, but Brown managed to substitute the slip for a winning betting slip on the same race. This suggests that they had placed a bet on each horse in the race and chosen to write off the bookie's edge as part of the costs of production.)

In John Allen Paulos' book *Innumeracy: The Stock Market Scam*, he points out that, while this is an illegal con if performed deliberately, there is a degree to which any group of predictions can result in a similar outcome. For instance, out of a group of sincere publishers of stock newsletters, or quack medicine sellers, or television evangelists, there will at any point be a proportion who have made successful predictions, or sold cures to people who were already on the point of recovery. Human irrationality being what it is, there is a tendency for these individuals to be given far too much credit for their foresight.

Indeed, the whole financial sector tends to be prone to making this mistake. As Nassim Nicholas Taleb has pointed out in *Fooled by Randomness*, the individuals who make disastrous losses tend to leave the industry, while those who survive are the ones who have mostly managed to succeed, or at least not to fail too badly. But sheer luck will account for a high proportion of the variation between the winners and losers. The result is that survivors have far too high an opinion of their skill and far too little regard for the element of luck that they have enjoyed. You could, for instance, apply this logic to the hedge fund industry – traders with a few years of good results are often seen as gurus and entrusted with huge sums of money, but a huge proportion of such funds end up failing nonetheless. Derren Brown argued in *The System* that the same bias affects those who believe in spiritualism or homeopathy.

At times, this tendency to trust that past results will

lead to future success is exploited more explicitly. The 'algorithm scam' is one where the conman fabricates past success by claiming to have made trades that would indeed have been successful. The claim is that they are a trader with a secret algorithm that allows for a high rate of success in the markets. The exact way the algorithm works is, of course, proprietary information. This scam has been used to lure investors to a variety of fates – in some cases the money has disappeared. In others, the con artists simply invested their capital in an index fund, extracted large fees for modest returns, and sat tight until investors lost patience.

Of course there really are quants and funds out there with extraordinary algorithms, some of which are highly profitable (for instance, see Chapter 6). So the trick is to ignore the advertising and work out what is really going on. The ones who shout the loudest about their algorithm may well be the conmen.

Ponzi Schemes and Pyramids

Bernie Madoff's notorious fund relied on a version of the algorithm scam to entice investors – in his case the fabrication of fabulous past results encouraged investors into the investment management and advisory division of his well-known Wall Street firm. However, his criminality went a stage further: the entire investment division was a Ponzi scheme, which collapsed during the 2008 financial crisis.

Ponzi schemes are named after Charles Ponzi, an Italian-American who came to notoriety in the early 1920s for his fraudulent moneymaking scheme. He claimed to be running an arbitrage business whereby you could buy discounted postal reply coupons abroad and redeem them at full value in the USA. He promised investors 50% profit within 45 days, or 100% profit within 90 days. However, he was paying out the old investors' profits from the money he took in from new investors and the scheme inevitably ended in disaster.

Madoff was running a similar scam. The early profits made by investors in his fund encouraged others to join them. He was able to keep paying out healthy profits over a long period because earlier investors kept reinvesting and there was a constant flow of new investors. Madoff himself was astonished that he was allowed to get away with this for so long – many of the trades he claimed to have made would have been easy to disprove. When the scheme finally collapsed in 2008, there was a black hole of £65 billion, consisting of investors' lost funds and profits they had been promised.

Pyramid schemes operate in a similar manner to Ponzi schemes, though unlike a Ponzi they aren't completely run by the originator – investors are recruited by the people participating in the scheme, which often relies on selling bogus products. For instance, you are offered the chance to join a wonderful moneymaking scheme in which you will be selling health supplements. The only catch is that there is little profit in the sale of the supplements, and your actual payments are mostly dependent on you recruiting more people to the scheme.

This is the crucial part of a pyramid. Each person who joins is given the task of recruiting several new members. The new members pay a joining fee or similar, part of which is passed up the pyramid to the founders of the scheme. Early joiners can indeed do well out of this, as they find it relatively easy to find new recruits, and earn the promised money.

The catch comes when the basic mathematics of the scheme is examined. Say that each member has to recruit four new members in order to receive their payment. At the first level you have the founder of the scheme. The second level has four new recruits, the third level has sixteen, the fourth has sixty-four and so on (see Figure 22). The number of people needed to reach the fifteenth layer of this scheme would be 1,000 million, three times the number of people who live in the USA.

As a result pyramid schemes tend to collapse relatively quickly. Ponzi schemes can be surprisingly longlasting (since early investors can reinvest), so there are an increasing number of pyramid schemes that look for ways to allow people to make repeat payments, in order to sustain the illusion for longer. But in the end Ponzis and pyramids have the same fundamental flaw, which is that they are redistributing wealth (to the originators) rather than generating it.

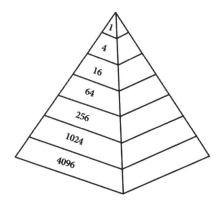

Figure 22. The mathematics of the pyramid scheme.

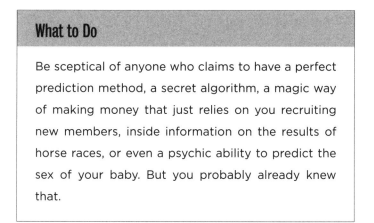

What to Do

Be sceptical of anyone who claims to have a perfect prediction method, a secret algorithm, a magic way of making money that just relies on you recruiting new members, inside information on the results of horse races, or even a psychic ability to predict the sex of your baby. But you probably already knew that.

Cracking the Lottery Code

We have seen that lotteries (and the associated scratchcards) have a particularly low expected return among gambling methods, often returning as little as 50% of your original investment. The fact that they have high potential payouts attracts a lot of gamblers from relatively poor backgrounds, which is why they are often seen as a way of taxing the poor. However, at least one man has found a genuine way to beat the lottery odds.

Mohan Srivastava, a statistician from Toronto, became fascinated by the problem of how to identify winning scratchcards after a colleague gave him a few as a joke present. His work for mining companies involved trying to find ways to extract reliable statistical information from limited amounts of data. Looking at the scratchcards he pondered the fact that the cards are mass-produced, under the requirement of producing exactly the right number of winning cards. So clearly rather than the random generation of numbers, there was a pseudo-random process at work in the production of the cards, which meant there might be a flaw in the system.

The specific cards he was considering displayed games of tic-tac-toe (equivalent to the UK-game of noughts and crosses). The tic-tac-toe boards started out printed with seemingly random numbers. At the side of the card a panel of winning numbers was revealed when you scratched off the covering. If you got a row or column of winning numbers among the tic-tac-toe games, you would get the prize indicated. There were eight tic-tac-toe games on each card, theoretically giving you eight chances to win.

The advantage of a mathematical brain is that it is trained to look for genuine patterns, and tends to be fairly patient and persistent. Srivastava tried a succession of theories as to what patterns might be relevant – it took him a while to spot the actual pattern, but once he did it was a laughably simple flaw. On each card some of the numbers on the tic-tac-toe boards are repeated and some aren't. If you mark up only the singletons, then when there are three of these in a row or column you have a winning card (see Figure 23).

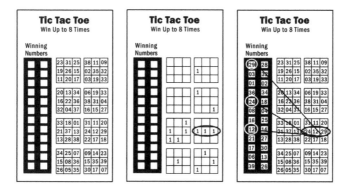

Figure 23. To find the winning cards, Srivistava marked up a grid as in the second image – a blank indicates a repeated number, a 1 indicates a singleton. The row of singletons indicates a prizewinner.

Srivastava did hesitate over whether to take advantage of the discovery, but he calculated that it really wasn't worth the time involved in driving around collecting winning cards – he might make a few hundred dollars a week, which was not as much as his more interesting job as a statistician. So instead he took his findings to the lottery board. Initially they refused to meet him, thinking he was a crank, but eventually he proved to them there was a flaw in the system by sending a package of scratchcards with the winning ones identified, and they finally withdrew that game from sale.

They claimed that this was a freak, and that in general scratchcards are unbreakable – the lottery companies are audited to check that this is the case. However, Srivastava went on to find similar flaws in cards in several other parts of the world, which would increase the chance of picking a winning card by up to anything from 30% to 100%.

There's no clear evidence that anyone has used such anomalies to take huge profits from a game, but there is some anecdotal evidence that the flaws in other games may have been exploited. The Massachusetts state auditor reported on the local lottery and found that, for instance, one unnamed person cashed in 1,588 winning tickets between 2002 and 2004, taking a total of $2.84 million. (The report does not provide the name of the lucky winner.) A 1999 audit showed that the ten top winners had won 842 times, taking a total of $1.8 million. These kinds of concentration of winnings among a small group of people are extremely unlikely to have happened by mere chance.

There is a possibility that such freak results are down to criminal activity. Lottery tickets don't have a high payout, but they have been used in some areas to launder the proceeds of crime. Criminals work on the assumption that having proof of winning lottery tickets for 50–60% of their original money will give them an alibi. However, it is at least possible that some of these anomalies are down to individuals discovering ways of decoding the information displayed on the tickets and identifying winners in advance. After all, not everyone is as honest as Mohan Srivastava.

Winning the Lottery

In most gambling situations, the odds are rigged so that you can't simply bet on every option and guarantee a profit. However, when it comes to lotteries there are occasionally situations (for instance where the jackpot hasn't been won for a few draws) where it is theoretically possible to buy every single ticket and be sure of winning.

This idea has occurred to a lot of people, although a few moments' thought suggest that logistically it would be extremely complex. However, given enough determination or resources, it is indeed possible – which we know because it has actually happened a few times. In the early 1960s, the Romanian economist Stefan Mandel was looking for a way to emigrate from his native country. In the age before personal computers it was a fearsome task simply to calculate the odds and to identify whether the feat was feasible. Also Mandel was unable to raise the finances to buy every single ticket – so instead he manually worked out the combinations he required to guarantee getting at least five of the six numbers. He was luckier than that though – he managed to hit the jackpot first time and raised enough money to migrate with his family to Australia.

There, he continued to ponder the logistics of a more ambitious attempt. To this end he organized a syndicate which could raise enough money to fund an attempt. Over the next few decades his syndicate managed to win the lottery 12 times in Australia. Stefan Klincewicz, who knows Mandel and organized a similar heist on the Irish lottery in 1992, has explained the difficulties faced by any syndicate attempting this feat. His own attempt involved raising the money, setting up offices from which the ticket-buying could be supervised, and recruiting buyers by word of mouth, as well as people to fill in the correct numbers. The lottery slips needed to be collected in advance from various sources, and hotel rooms booked at various locations to spread out the purchases geographically. They looked for quiet purchase points where they wouldn't be too conspicuous, and finally brought all the tickets to a secure central location. All of this needs to be paid for, so you need a lottery with sufficient guaranteed profit to cover considerable costs as well.

In the case of Klincewicz's attempt, the Irish lottery spotted that something was going on and restricted volume ticket sales days before the draw, as a result of which they were only able to buy 88% of the tickets. On the one hand they were lucky, as the winning ticket was nonetheless among their purchases. However, they were unlucky in that they had to share that jackpot with two other winners, which is one of the risks this method runs. The syndicate still turned a profit, once all the smaller prizes were added to the jackpot. But it was significantly smaller than it might have been.

By this point, Mandel's syndicate had long since come up against obstacles in the form of laws in Australia passed specifically to make syndicate lottery betting more difficult. He bided his time and identified another potential winning lottery, in the US state of Virginia in 1992. It was an even more complex operation as his syndicate had to buy the tickets through long-distance negotiations with local stores in the US. However, they were successful in their attempt, winning the $28 million top prize and 135,000 smaller prizes. Mandel eventually retired to a small island in the South Pacific, a wealthy man.

It's worth knowing that many lotteries have rules against such mass purchases and will suspend or restrict sales where they suspect such an attempt is underway. Also the prize structures are often designed to make it impossible to guarantee a profit, since high prizes can be shared between several winners. So while Mandel is an intriguing figure, it may be that his success and those of a few other syndicates around the world have turned this particular moneymaking scheme into a dead-end.

Enhance Your Chances

Lotteries aren't the only game of chance that have been beaten by cunning punters. The American game show *Press Your Luck* was the scene for an extraordinary winning run, which was down to mathematical pattern recognition rather than luck.

Michael Larson grew up in Ohio, and developed a fascination with loopholes and shortcuts which would allow him to gain easy money. For instance, as a young man, he opened multiple bank accounts to take advantage of a free cash giveaway for new account holders. He didn't hold down many long-term jobs, and became obsessed with watching television to find ways to earn money from game shows or infomercials.

Press Your Luck was a relative newcomer to the television, having first aired in 1983, the previous year. Larson spent days watching tapes of the show and managed to identify a weakness. The contestants won money by first answering trivia questions, which then earned them spins on a computerized board of flashing lights. Pressing a red button stopped the sequence on one of 18 squares, each of which had three possible outcomes. Nine of the 54 outcomes were 'whammies' which meant you lost any money you had accumulated. So in theory you had odds of 1 in 6 on losing everything. As a result, your odds of losing everything were over 50% by the fourth spin of the board – at each spin the odds of winning something were 5/6 so the odds of winning four in a row was ($5/6 \times 5/6 \times 5/6 \times 5/6 = 625/1,296$). Once you had accumulated enough money you could pass your spins on to other contestants to try to keep your gains.

The board had been designed to mimic a random sequence, with producers having tested the system and assured themselves that the biggest payouts would be no more than $25,000. In theory a run of spins could go on indefinitely, as certain outcomes gave you an additional spin, but the producers didn't expect anyone to be able to hit these squares reliably.

Larson realized that the pseudo-random sequence was actually a fairly simple repetition of five separate shorter sequences. And there were two squares among the 18 for which you would never get a whammy and would always win an extra spin. Having laboriously memorized this pattern, he made his way to the CBS studios and managed to persuade them to allow him to be a contestant in a show that was taped on 19 May 1984.

He started badly, getting a trivia question wrong and getting a whammy on his first attempt. However, he gradually got the hang of the board in the live situation and started managing to hit his intended squares more often. (He also needed a bit of luck as he missed the whammy on the other times he failed.) Finally he hit a long run of success, hitting his destination squares 45 times in a row. Producers watched in horror and an increasing state of panic, while the game show host Peter Tomarken became increasingly rattled by what he was witnessing. The eventual show went on for so long it had to be split into two episodes when it was screened (and CBS was so horrified they refused to repeat it for years). The show is fascinating to watch as you can see that Larson celebrates his success as soon he hits each square, **before** the prize has been revealed. He ended up winning $110,237, plus a boat and two all-inclusive holidays before deciding to cash in his chips and passing on his last few spins.

The producers examined the tapes of the show, hoping to prove he had cheated. The odds on him getting 45 winning spins in a row were $(5/6)^{45}$ = 0.027%, so it was highly unlikely to have happened by chance. It seemed clear enough that Larson had somehow figured out the sequence (indeed the producers later admitted this possibility had been discussed and dismissed in early show meetings). But he hadn't broken anything in the rules or conditions, so they reluctantly paid up. And then they immediately reworked their board with a much more complex sequence driven by a more powerful computer.

Sad to say, Larson's career thereafter was patchy. He lost a significant amount of the money pursuing other long-shot ideas, had another part of it stolen in a house robbery, and ended up running a Ponzi scheme in Ohio in the 1990s, selling shares in a non-existent lottery and swindling 20,000 investors to the tune of $3 million. He died of cancer while on the run from the authorities. But his appearance on *Press Your Luck* remains legendary and has inspired many people to hunt for loopholes in game shows and even to try to cheat their way to the big prize.

The Hole-in-One Gang (and Unreliable Intuitions)

As we've seen, our intuitions on probability can be remarkably inaccurate. The Monty Hall Problem is another great example. In the 1960s US game show *Let's Make a Deal* the host Monty Hall would offer contestants the following deal:

There is a car behind one of three doors, but a goat behind the other two. You are asked to choose one of the doors (say it's door 1). Then the host will open one other door (door 3) to show you a goat. Now the host offers you a choice: stick with door 1, or switch to the remaining door 2. To win the car, should you switch or not?

Figure 24. The Monty Hall Problem.

Most people's first instinct is that there is no reason to switch. There are two closed doors, so the odds on each being the door with the car behind it seem to be 50% (see Figure 24).

In fact you should switch because the actual odds are 33.33...% that the door you initially chose has the car behind it, and 66.66...% that the other remaining door does. When this problem was first published with the correct answer in a US magazine, thousands of people complained that the solution was wrong, including several maths professors, so clearly it is a much more difficult problem for us to understand than it looks.

The best way to understand why the odds aren't equal for the two closed doors is to imagine the same situation occurring again, but now with 100 doors concealing 99 goats and 1 car. After you choose door 1, the host opens 98 doors with goats behind them, leaving your original choice and door 100 (see Figure 25).

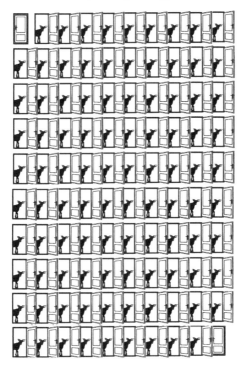

Figure 25. The 100 door version of the Monty Hall Problem.

In this version it is easier to understand that your original choice had just a 1% chance of being correct, so the other remaining door must have a 99 in 100 chance of being the one with the car. By ruling out 98 of the other 99 doors, the host has given you almost certain information that the remaining door is the correct one – because the only scenario in which switching would be wrong is if you happened to have chosen the correct door the first time, which would only have happened 1 in 100 times. The logic for the original scenario with three doors is exactly the same: your original choice had a 1 in 3 chance and that hasn't changed, so the other door has a 2 in 3 chance and you should switch.

The Birthday Conundrum

Another counterintuitive statistical quirk arises when you consider the chances of two people in a random group of people sharing a birthday. Most people are surprised to discover that for a group of people of 23 or more, the odds on this happening are better than 50%.*

Again, we can understand this more clearly by unwrapping the maths. Let's work out the odds of each person in turn **not** sharing a birthday with anyone else.

* This is assuming that birthdays are randomly distributed through the year and ignoring leap years, twins and other minor anomalies.

For one person the odds are 365/365 as there is not yet anyone they could share with.

For the second person, the odds are 364/365 (because the one remaining day is the birthday of person 1).

For the third person the odds are 363/365.

And so on until for person number 23 the odds are 343/365.

In order to find the odds of all of these events being true (that is, no-one sharing a birthday) we multiply the odds together:

$$365/365 \times 364/365 \times 363/365 \times ... \times 343/365$$
$$= 0.493$$

So there is a 49.3% chance of **no-one** in a group of 23 people sharing a birthday, meaning that in 50.7% of such groups there will be at least two people who do share a birthday.

The difficulty we have in intuitively seeing the correct odds in many situations can create value betting opportunities, if the other person is gambling based on their intuition rather than the genuine odds. Of course bookmakers are pretty smart and spend a lot of time analysing the statistics in order to avoid obvious mistakes.

However, there are still situations in which they can slip up. A great example of this comes from the Hole-in-One gang.

In 1991 Paul Simmons and John Carter, both of whom were from Essex in the UK and had worked in the betting industry, spotted a potential money-spinner. Many bookmakers treated bets on holes-in-one at golf tournaments as a novelty long-odds bet, and the odds offered varied widely. After studying the exact statistics over decades of major golf tournaments, the pair concluded that the average actual odds for any player to score a hole-in-one at any given tournament were better than even, but they had seen odds varying from 3/1 on this outcome up to 100/1 for accumulators on the same thing happening at several tournaments.

For an example of how the real odds work for this bet, consider the US Open. Statistically it takes professional golfers about 3,000 plays of a short par-three hole to bring about a single hole-in-one, which does make it look like an outrageously unlikely event if you view it in isolation. However, the US Open will feature 156 players and about three or four par-three holes, and this immediately shortens the odds. For instance if there are four par-threes, we can calculate the number of times each hole will be played thus:

156 players play the first two days: 156 × 4 holes × 2
= 1,248

After the cut the field is whittled down to a minimum of 60 players, who will each play these holes two more times:

60 × 4 holes × 2 = 480

In all there will be a minimum of 1,728 attempts at the short holes, so given a probability of 1 in 3,000 for each attempt, we have a total probability of at least $\frac{1,728}{3,000}$ (= 57.6%) for there to be at least one hole-in-one.

The pair came up with similar estimates of the odds for a range of forthcoming tournaments. Thus far, this was straightforward statistical analysis rather than any kind of a heist or con. But when it came to placing the bets, a bit of subterfuge was required. Simmons and Carter knew that the major bookmakers offered accurate odds on this bet, so in order to take advantage of their knowledge they had to place bets at a large number of smaller bookmakers, each of whom was mispricing the risk to some extent. This was the pre-internet era, so they embarked on a whirlwind road trip of towns around Britain, placing bet after bet on the individual tournaments they had chosen to target (the US and UK Open, Benson and Hedges, PGA and European Open) and on multiple outcomes of two, three, four or five of the tournaments featuring holes-in-one.

In the event there were holes-in-one at all of those tournaments that year, and the two men had notional winnings of £500,000. At least one bookmaker closed down and left the country rather than pay, and a few others refused to honour the bets, arguing that they had been unfairly swindled. But the majority of the bookmakers kept their word, so Simmons and Carter came out of it with a very healthy profit.

Of course, bookmakers around the world have all heard this story and it would be next to impossible to find one offering such inaccurate odds on a hole-in-one these days. However, the moral of the story is that where people are pricing risk based on their intuition rather than the statistics, there will often be an opportunity to profit from the resulting errors.

Good and Bad Information

In any kind of gambling or investment, it is theoretically possible to improve your edge by using information or analysis that other people don't have access to. This is not something I recommend, of course, but one way of rigging the odds in your favour is to have inside information. Insider trading in many forms of business is illegal, for good reason. When insiders know information that will later become available to others, they are able to gain an unfair market advantage in the short term. There is some debate as to whether this really should be illegal – some make the comparison with other markets in which information can be asymmetric and argue that the sooner new information is applied to a market the sooner the price will adjust. However, let's just accept for now that it is illegal and move on.

Another (now illegal) way of manipulating the market is 'spoofing' in which traders place large orders for financial instruments such as futures, only to cancel them before the purchase is completed. This can trigger sharp, temporary movements in the price, which can be separately used to make a profit. In 2015, the US Department of Justice charged Navinder Singh Sarao (also known as the Hounslow day-trader) with a variety of criminal counts, including the use of spoofing algorithms, which they claimed had led to the 2010 Flash Crash, in which millions of orders were made, then modified or cancelled in a short period of time.

Another, more legal, way of gaining an edge in investment terms is to find improved ways to analyse information. In the next section we will look at the work of quants and high-frequency traders, whose operation relies heavily on finding algorithms and models that give them even the most fleeting of advantages.

In general, if you can find legitimate ways to analyse the information that are superior to those who are setting the odds you can, in theory, beat the bookies or the markets consistently. For instance Steven Skiena is a maths professor who worked out a method for predicting the results of sports matches with 65% accuracy and thus for beating the bookmaker's odds. Unfortunately, his most successful system, known as Maven, was developed for the peculiarly obscure sport of jai alai (a version of pelota, in which a ball is propelled with handheld devices and bounced off a wall), and would not be much use to anyone now that he has publicly explained the system – betting on jai alai follows the pari-mutuel system in which all similar bets are pooled, the bookmaker's cut taken and the winnings shared out, so a system that is well known will dilute any potential winnings. However, Skiena's book *Calculated Bets* is a fascinating and humorous insight into the process of developing a genuinely successful betting system and the algorithms used. (Skiena is also an expert on the mathematical development of algorithms in general.) One of the methods he describes in the book uses **Monte Carlo methods**, whereby large random samples are used to try to model features of a system that might be susceptible to deterministic analysis (in which you can treat events as independent rather than as parts of a sequence). A well-known example is that you can come up with a pretty good approximation of pi by drawing a circle in a square (where the circle touches the edges of the square at a tangent). If you place this on a floor and scatter a large number of small objects randomly over the whole thing, then measure the ratio of the number of objects in the circle to the total number of objects in the square the result will be pretty close to a quarter of pi. The fact that we do not have to actually calculate this using traditional methods is one of the strengths of Monte Carlo analysis.

Skiena had worked on a variety of betting systems since he was young, saying that this, along with his fascination with baseball statistics, was his inspiration for becoming a mathematician. However, Maven was developed in co-operation with his graduate students and the proceeds were donated to his university. His personal observation on an area of gambling that might be susceptible to a really good betting system in the future is the concept of poker-playing bots (computer programs) that could play online. These have since been delivered – the evidence thus far is that poker-playing bots can, when well programmed, outplay many humans, partly because of the consistency with which they make correct decisions and their immunity to emotion. However, they are just as prone to being undone by variance as human players and are not a reliable way to make money.

Like Skiena, the pollster Nate Silver's mathematical expertise is inextricably linked with his love of sports. Silver's reputation for making brilliant political predictions was originally rooted in his analysis of baseball odds. While working as an economic consultant in the early 1990s, Silver developed PECOTA (Player Empirical Comparison and Optimization Test Algorithm), a statistical system that projects the future performance of hitters and pitchers. Silver's original approach to this task could easily serve as a model for how to develop your own new methods of processing information rather than relying on existing analysis.

One of his innovations was that, unlike other baseball projection systems, PECOTA tagged each player against a group of 'comparable' players and their past performance patterns. He also incorporated unique sources of information into his system, for instance figures from a handheld radar gun giving the speed at which a pitcher released the ball. Silver also emphasizes the importance of giving results as a range of probabilities rather than as a single-figure projection of what the actual result will be. For instance when he moved over to political analysis, rather than saying something like 'Obama has an 8% lead in North Carolina', he would prefer to say 'Obama has a 62% chance of victory in North Carolina.'

The eighteenth-century minister Thomas Bayes introduced Bayesian analysis as a way of constantly updating the probability assigned to a hypothesis as more evidence is assessed. Silver harks back to this kind of analysis in one important respect – by emphasizing how uncertain each individual statistical tool at his disposal is, he suggests constantly testing actual results against the model and calibrating it so that the model gradually comes into line with reality. This is a standard statistical approach, but the use of probabilistic outputs from the PECOTA model gave it a new relevance.

In his book *The Signal and the Noise*, Silver discusses the use of models in sport, politics, weather forecasting and finance, among other topics. He moved into political projection in the 2000s, initially with much success. His predictions for the 2008 and 2012 US presidential elections were, for instance, among the closest to the actual results.

For all Silver's strengths he got the 2016 result wrong. To be fair, his projection that Trump had only a 30% chance of winning can be defended as being closer than some pundits, but it gave some people a false sense that a Clinton victory was inevitable. He was not alone: there has been a sequence of elections in recent years in which all the polls have been misleading. So why do pollsters get things so wrong and can we learn any lessons from their failings?

When the Polls Are Wrong

The art of opinion polling is a pretty complex business. The aim is to provide a scientific survey of the views of a group of people. In order for this to have any validity it needs to be based on a sufficiently large proportion of the chosen group – as with any set of data, a small sample is more likely to suffer from misleading biases. The first thing to look at when considering whether a poll is valid is whether it has been based on sufficient responses. Anything less than a few hundred responses is likely to be inaccurate. But a large sample is insufficient for avoiding error – in the 1936 US Presidential Election, *Literary Digest* magazine sent out 10 million postcards and concluded from the 2.3 million that were returned that Alfred Landon was leading Franklin Roosevelt by 57 to 43%. Crucially, they hadn't made any effort to check whether these responses came from a biased sample. The reputation of modern opinion polling methods could be said to date from their error – for the same election, George Gallup, a young pollster, relied on a sample of only 50,000 but used information about the respondents to weight their responses. As a result, he made the accurate prediction that Roosevelt would win by a landslide.

The problem is how to make sure that the sample is representative of the group. Polling companies use a variety of methods to achieve this – for instance they can use **random sampling** of a large proportion of the population to minimize the danger of bias (as opposed to the *Literary Digest* poll, which was too reliant on the single demographic of their readers). Or they can use **quota sampling** in which they attempt to ensure that their sample has, for instance, a good balance of people of different gender and from different age groups.

Finally polling companies use a variety of adjustments to calibrate their results. For instance, it's hard to find a truly representative sample – often the group will be slightly biased. If a poll about an American state election is based on a sample of 50% each of registered Democrats and Republicans, and the actual population has 52% of Democrats and 48% of Republicans, the polling company might choose to weight every Democrat response by 104% and every Republican response by 96% in order to balance out their figures.

The next problem is that even for a representative sample we have to allow for a margin of error. You can use basic standard deviation calculations that show that the

margin of error for a poll of 1,000 people will tend to be 3% in either direction. This means that a result showing the group divided 50/50 on a subject might indicate an actual split of 53/47 in either direction, with a gap of 6% between the upper and lower limits. People often mistakenly describe a poll as being 'wrong' when it was within the margin of error. (This is one reason why Nate Silver believes in giving a range of percentage likelihoods rather than a specific prediction of the final result in elections.)

Finally it's worth bearing in mind that there are many more technicalities about polling, for instance how the questions were phrased, what order they were placed in, and so on, all of which can affect people's responses. Polling companies try to allow for this by using the same questions and formats over time and calibrating their adjustments to the actual outcomes. An example of this is that there is a particularly large disparity between how many young people say they are going to vote in an election and how many actually turn out. So companies will weight their responses downward accordingly, and will usually rely on actual figures from recent elections for their estimate of how to weight these responses.

So why do polling companies still get things wrong? Recently the Brexit referendum, President Trump's 2016 victory and the UK 2017 General Election are notable examples of elections where the polling companies performed badly, even allowing for the margin of error.

Survation were the only UK pollster to predict Brexit while Nate Silver's prediction that Trump had only a 30% chance was much closer than many other companies who confidently predicted a Clinton victory.

The reasons are many and complex, but a key thing to notice in these elections is that major shifts in voting patterns played a significant part in the final result. Trump's victory and the Brexit vote were based on surprising turnout among demographics who had previously tended not to vote. There were also significant shifts in sectors of the population such as the rust belt states which pollsters discounted based on their experience of recent elections.

Nate Silver has argued that the polling errors in these elections weren't down to the data analysis, but to conventional polling techniques including the use of instinct and experience to discount apparent 'outliers' – this theory suggests that pollsters looked at the data and simply didn't believe it, so adjusted their conclusions to something closer to their somewhat elitist expectations in elections where anti-establishment figures were drawing the popular support.

There certainly is a tendency for polls to 'cluster' shortly before an election, meaning that the polling companies don't want to be the outlier that gets the prediction wrong. An example of this is the final poll carried out by the company Survation on the eve of the 2015 UK general election. This predicted a Conservative overall majority, while most polls were showing a hung parliament.

Damian Lyons Lowe, the founder and CEO of Survation, has admitted that they 'chickened out' of publishing this poll, feeling it must be a freak result as it was so far from the average of other polls. (They learned their lesson from this, and listened to the data in 2016 when they correctly predicted Brexit.)

So there are non-mathematical reasons for the polling errors. But Nate Silver's attempt to deflect the criticism of data analysis is disingenuous and to some extent unnecessary. The truth is that when you are in a period of rapid change in politics, the methods polling companies use to adjust their samples can't be perfect. If, for instance, the turnout among specific groups of the population changes rapidly between two elections, the calibrations made by the polling companies will be prone to error. There is no perfect mathematical way to fix this problem, as polling companies are always having to 'fight the last war'.

This doesn't mean that polling is useless. It can still give us genuine insights into attitudes. But it is worth bearing in mind when it comes to assessing both political polls and any market research that a company may rely on – at the very least we should always note that any result within the pollsters' margin of error should be treated as a possible outcome. And if the surrounding circumstances suggest a period of rapidly changing attitudes to politics or to a type of product being researched, then the polls will be at their least reliable.

For gamblers, it's worth noting that the odds available on both Brexit and Trump victories were pretty generous. An unknown gambler reportedly won $620,000 on a $250,000 bet on Trump, while one woman in London who had never placed a bet before gambled £10,000 on a Brexit win at odds of 11/4, winning £27,500 in the process. But it's not just gamblers who can lose or gain from such results. Carlos Slim, the fifth wealthiest person in the world, is reputed to have lost roughly $5 billion in the immediate aftermath of the Trump election, simply because the value of the Mexican peso fell so rapidly.

Beyond the lessons that gamblers and investors can take from the uncertainty of the polling process, it also worth noting that in times of rapid change in the marketplace or electorate, we won't always be able to solve the problem of finding out what people 'really think'. Instead, we will sometimes need to just go out on a limb and take a risk. We can nonetheless learn from Nate Silver and calibrate our approach as we move along. Launching a business or a new line of products is never easy – and all the market research and advance planning in the world can't protect us from getting things wrong from time to time. The main thing is to keep trying and to keep fine-tuning our approach based on the feedback until we find a winning formula.

Chapter 5 Summary

1. Many hacks and cracks rely on spotting a mathematical anomaly or recognizing a pattern that other people may have missed.

2. In this chapter we've seen stories about people who managed to find ways to beat lotteries, game shows, casinos and bookmakers. The specific examples given are well known to modern practitioners – but the stories nonetheless can be used as inspiration that there might be a loophole or opportunity still out there.

3. Most hacks and cracks are easier to imagine than to actually put into practice.

4. Beware of con artists whose offers and promises seem too good to be true.

5. All forms of data analysis and systems have their flaws but you can gradually calibrate your approach based on results to reach better-informed decisions.

Designing the Next Google

The biggest risk is not taking any risk... In a world that
is changing really quickly the only strategy that is
guaranteed to fail is not taking risks.
Mark Zuckerberg

Algorithms and solutions to complex maths puzzles have
underpinned some of the most successful start-up tech
businesses of recent years and have also come to dominate
the financial sector. In this chapter we'll look at some of the
basics of how mathematical modelling is transforming the
world of finance and business. Some of the mathematics
referenced in this chapter is quite advanced. Rather than
get bogged down trying to explain the most complex
problems, I'll mainly aim to give an overview that gives an
indication of the kind of methods involved. On the other
hand some of the real giants of the industry have been
built on surprisingly modest mathematical roots.

Google and the Matrix

Sergey Brin and Larry Page, the founders of the tech giant Google, met in 1995 when Page visited Stanford University for a recruitment weekend and Brin, a graduate student, helped to show him around. At first they didn't get along too well, but once Page was studying at Stanford the two started to collaborate on a project, which was called BackRub. It was an attempt to find a way to navigate and document the structure of the web which would eventually be renamed PageRank. This would go on to become a breakthrough in internet-search-engine technology, and remains at the heart of Google's workings today.

The earliest search engines simply ranked pages on how many times a given search term occurred on a page. This could mean that, for instance, a repetitively written page which gave a list of names of types of roses could easily outrank the most informative, useful webpage written by the world authority on rose taxonomy, simply because of the word count. The fixes to the problems that this created in predecessors such as Altavista were still not producing particularly good results. The fundamental insight that Brin and Page applied to PageRank was that you could rank a page based on not only its relevance to your query, but also on how many other pages linked to that page.

One way of thinking about this problem was to imagine a 'random walk', in which a web user followed completely random links around the internet. The probability of this user ending up on a given page would be one way of assessing its importance.

Of course there are over 25 billion pages on the internet, so how would you even start to calculate this? The solution Brin and Page came up with was, at heart, based on surprisingly elementary mathematics.

Let's visualize the internet as an extremely simple network represented as a directed graph in which each page is a node, and the arrows represent the links between the nodes (see Figure 26).

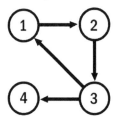

Figure 26. A simple network with numbered nodes representing internet pages.

Next, if we assume that each link made by the user is random, then each page to which they could link has equal probability of being the target. Now we can form a matrix (a rectangular array of numbers for which functions such as addition and multiplication are defined) of the probabilities of linking between the pages. From node 1 we can only link to node 2, so we assign that step 100% probability (= 1) in the first row, second column. From node 3 there is a 50% chance (= $\frac{1}{2}$) of linking to nodes 1 or

4, thus the third row has $\frac{1}{2}$ in the first and fourth column. This gives the following matrix to represent this network:

$$\begin{pmatrix} 0 & 1 & 0 & 0 \\ 0 & 0 & 1 & 0 \\ \frac{1}{2} & 0 & 0 & \frac{1}{2} \\ 0 & 0 & 0 & 0 \end{pmatrix}$$

There is a slight problem here, which is that from node 4 it is impossible to link to anywhere. As a result it is called a **dangling** node. One way we can deal with this is by assuming that from node 4 the web user will randomly choose any page (and thus we assign each page $\frac{1}{4}$ probability in the fourth row):

$$\begin{pmatrix} 0 & 1 & 0 & 0 \\ 0 & 0 & 1 & 0 \\ \frac{1}{2} & 0 & 0 & \frac{1}{2} \\ \frac{1}{4} & \frac{1}{4} & \frac{1}{4} & \frac{1}{4} \end{pmatrix}$$

Alternatively we can assume that at any point there is something from a 1% to 15% chance of the user choosing a random page rather than following a link – in their original paper on the subject Brin and Page used a **damping factor** of 0.85 to deal with this complication (with $1 - 0.85 = 0.15$

being the probability of the user choosing a web page by means other than following a link).

This way of proceeding allowed them in theory a way of assigning a rank to every result a search throws up – the current PageRank gives a fairly simple ranking of one to ten, which seems to be based on a logarithmic scale, with (for instance) numbers 1–10 being assigned ten, 11–100 being assigned nine, 101–1000 being assigned eight, and so on.

So now all they had to do was find a way of doing this calculation for a matrix of 25 billion pages... And there is an additional problem, which is that the only way you can calculate the PageRank of any one page is if you already know the PageRanks of all the other pages it links to.

Early in a university maths course you would be likely to encounter the subject of calculating eigenvectors and eigenvalues for a matrix, and this was the underlying principle that they used at this point. This is tricky maths, of course, but not super-difficult. One way of calculating the solution to an eigensystem is to use the power method. Without getting into too much detail, this essentially involves starting off with a reasonable estimate of the probability for each page, and then reiterating a function that forces the results to converge more and more closely to the actual figure. Once the results are as accurate as possible, this function will keep feeding back more or less the same results each time.

And, in practice, this is what PageRank did from the

start. Once it has been up and running for a while, each round of calculations can start from the previous set of results, so only new pages start out with the 'guesstimate' initial figure, and the accuracy of the existing results allows for the results to be calibrated rapidly and constantly.

The precise algorithms that go into Google and some of the adjustments that have been made to weed out erroneous results and manipulations are of course more complex (and mostly secret). And a lot of the mathematics that has gone into the company's operations in the past two decades are highly complex. But right at the heart of the search engine lies PageRank, for which the spark of inspiration came from first-year college maths ideas. And it may yet be that the inspiration for building the 'next Google' will come from a similarly humble piece of mathematical thinking.

The Maths of Facebook

Algorithms are a set of rules or operations to be used in problem-solving or calculations. They have become a huge part of our daily lives, especially in the way we interact with the web. Internet businesses large and small rely on algorithms to target advertisements and content at consumers. The information, posts and adverts that pop up in your news feed on Facebook or timeline on Twitter are generated by algorithms. These use data sets based on a specific set of inputs that then generate automated decisions as outputs.

In some respects this can be quite sinister. We've seen in the Cambridge Analytica scandal how data from 50 million Facebook profiles was allegedly used without permission to target US voters with political feeds which were generated by their profile information. Cathy O'Neil, author of the brilliant book *Weapons of Math Destruction*, has described numerous other ways in which secret algorithms can affect public policy, your finances and the ways we are governed. For instance, sentencing in criminal cases in various countries around the world can sometimes depend on how 'high-risk' the individual involved is in terms of repeat offending. If they are deemed high-risk, then they are more likely to receive a custodial sentence. However, the algorithms involved in determining risk are far from perfect – and this is hard to detect since someone who has been in prison will have problems finding work when they come out, and will thus be at more risk of falling back into criminal behaviours, regardless of how 'high-risk' they really were. This is a case of an algorithm affecting what it is supposed to be measuring, and thus producing false feedback. Similarly, an autistic person who scores badly on an emotional intelligence test for an employer may be denied a job in which they would have performed well by the corporation's algorithms; the extra time without a regular income may feed back into their credit score; this may affect how the insurance algorithm prices their car insurance, and so on.

Like it or not, algorithms are here to stay, so the most important thing from a mathematical point of view is to understand how they work. Of course the exact algorithms are secret and change frequently as they are fine-tuned. But the overview of, for instance, Facebook's algorithm is that every single interaction you have with Facebook goes into the algorithm – every click, every like, every video you play, every advert you reject, every group you follow and so on. These are fed into the algorithm in order to sort the vast amount of information that is out there into order of 'relevancy'. The news feed then shows the items that have the highest relevancy, but continues to monitor how you respond to them in order to check if it has produced the right results. Every now and then Facebook tweak the algorithm, for instance to show you more posts from people who you have recently 'liked' and to exclude posts from people and organizations you haven't, meaning that your responses are creating an automatic filter.

Of course there is a well-documented danger here, which is that such algorithms lead to 'bubbles' in which people are only fed the information they will welcome. In the USA there is already a huge divide between conservative and liberal viewpoints, and one has to wonder whether this kind of algorithm-driven feed is making the problem worse by only supplying items that confirm the prejudices of people on either side of the division.

In some cases we can directly discover why particular items have been selected. On Facebook you can theoretically choose the 'Why Am I Seeing This?' option and discover which part of your profile has led to an advert being targeted at you. It will tell you, for instance, that it is because you are a forty-something man living in the London area (although this may in truth be only part of the actual process the algorithm has gone through). And if an algorithm seems to have given you rough treatment (for instance in applying for credit) you may sometimes be able to force the company to reveal the reasons behind their decision. It is even possible in some cases that the algorithms are behaving in a manner that would be discriminatory if practised in any other way, so it is worth challenging companies if you feel wronged. You may have difficulty getting a coherent response, however, as many algorithms (and artificial intelligence programs) are so opaque that even the programmers don't understand every nuance of how they operate.

Up to a point, understanding how the algorithms work can also affect how they treat you – if you use this knowledge when it comes to how you use a particular site. If, for instance, you are a conservative or liberal who feels that the 'filter bubble' is preventing you from seeing a wider spectrum of views, it might be wise to follow and click on a few of the news sources you would normally reject. This approach won't suit some people, who will only be angered by reading information that they disagree with. But if we are not careful the algorithms of the future are going to leave us in bubbles that we have consented to being imprisoned in. At the very least it is a good idea to try to get to grips with how pervasive they are becoming and to challenge them from time to time.

Encryption and the Banking System

Facebook and every other online corporation have to rely on high levels of security in the way they process information. The arts of cryptography and cryptanalysis (codemaking and codebreaking) have always appealed to the mathematically minded. Early examples such as the Caesar cipher relied on a simple mathematical function, replacing each letter with another from a set number of places along the alphabet. Such simple ciphers were breakable from the middle ages onwards – the Iraqi philosopher al-Kindi essentially kickstarted the art of codebreaking with his invention of frequency analysis, which relied on working out the relative frequency of each letter of the alphabet within the most common words. Using this system it became relatively easy to work out from the ciphertext which symbol represented each plaintext letter. When codemakers moved on to polyalphabetic codes such as the Bellaso cipher, and homophonic systems such as the Mantua cipher, more complex methods of analysis were required, but again they relied on mathematical methods. Similarly, it took Alan Turing and the brilliant mathematicians and academics of Bletchley Park to break the code used by the Germans' Enigma machines. In the process they took a big step towards the creation of modern computers, which have created an entirely new range of difficulties for those who want to encrypt messages, from the banking system to the creators of secure websites.

Since the late 1970s the best solution to this problem has come from **public-key encryption** (PKE), in which a key (which will generally be a series of numbers) which is accessible to all participants is combined with a private key that allows the intended recipient of the information to decode the message. For instance, consider the RSA (Rivest–Shamir–Adleman) encryption algorithm, the first widely used example of a PKE method, which is used by banks and secure commerce websites. It is based on the difficult task of finding the prime factors of very large numbers, especially large semi-primes which have only two very large prime factors. It is an asymmetric cryptographic algorithm, which means that there are two different keys, one of them can be given to everyone, while the other must be kept private. The public key is the semi-prime, while the private key is one of the prime factors. It is simple to multiply two large primes together, but much, much harder to factor large semi-primes quickly.

For instance, if you want to find the factors of 111, you can immediately tell this is a multiple of 3 (as the digits add up to a multiple of 3) and it is then easy to find that the other factor is 37, through division. However, for the number 2,183, a quick search for divisibility by 3, 5, 7, 11, ... will show you that these aren't factors, and it is only when you get to 37 that you can establish that this number is a multiple of 37 and 59. The higher the factors, the more complex this process becomes.

The danger is that mathematical shortcuts may in future make this a less arduous task, and RSA encryption may become much easier to break. This means that, for those who already have a good understanding of the problem of sieve theory (a technique to estimate the size of sifted sets of numbers), one potential area in which significant money can be made in future is by identifying new ways of factoring large semi-primes rapidly. But, be warned, this is a very difficult area, and one in which many advanced mathematicians are already hard at work.

Public-key encryption will almost certainly end up becoming redundant, for a good mathematical reason. Algorithms such as the quadratic sieve and the general number field sieve, which speed up the process of factorization, are already widely understood. One solution to this is to keep increasing the size of the numbers used as the keys. However, because these algorithms become more efficient as the numbers get bigger, the gap between the speed at which a processor can multiply two very large numbers together and the speed at which it can factorize one very large semi-prime is shrinking. This makes PKE gradually less powerful over time.

If you do have serious mathematical skills and want to get ahead of the next wave of cryptography, the most likely area of study is elliptic curve cryptography, which is in the early stages of adoption, but is already being used by sections of the US government, the Tor project and Bitcoin, for example. I can only take a very brief look at this here, as it is a difficult piece of mathematics, but here is an overview.

An elliptic curve is one that satisfies a particular mathematical equation:

$y^2 = x^3 + ax + b$

If you plot this as a graph, you get the kind of shape shown in Figure 27.

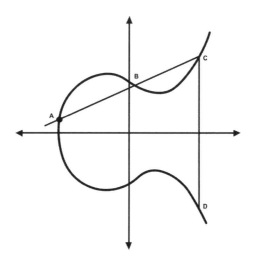

Figure 27. An elliptic curve.

This kind of curve has some particular properties that cryptographers can use.

Firstly it is horizontally symmetrical. Secondly, any non-vertical straight line will touch the curve in at most three places. Imagine a journey where you take any two points (A and B in the image) on the curve and draw a line through them to intersect the curve at one more place (C). Then you draw a line either straight up (if you are below the x-axis) or straight down (if you are above the x-axis) to the other side of the curve (D).

Any two points on a curve can thus be used to define a third and fourth point. The crucial thing for cryptographers is that, if you reiterate this process a few times, it is extremely hard to compute the starting point if you are only given the finishing point. And, crucially, the complexity of this task will keep on outrunning the difficulty of performing the calculations to find the path from the starting point to the finishing point. Also you can generate a high level of security from a fairly small key. This is vital because we have a proliferation of miniaturized devices which will find the increasing size of the keys used in PKE computationally impossible because they have insufficient computing power to deal with the multiplication itself.

There has been some controversy over one specific area of elliptic curve cryptography. The Dual Elliptic Curve Deterministic Random Bit Generator (Dual_EC_DRBG) is a random number generator promoted by American agencies, including the National Security Agency (NSA), as a standard system. However, there is widespread speculation that this generator may have a 'backdoor' which allows the NSA access to the encoded information. This doesn't mean that the method is inherently flawed, but it does mean that it is crucial for any device using it to have a strong random number emulator of its own (since the NSA pseudo-random one is theoretically 'backdoorable'). This is one area in which mathematicians are at work, trying to find new, more secure curves and methods that aren't reliant on the current standards.

The Byzantine Generals Problem and Bitcoin

Digital currencies of various sorts are becoming a vital part of the twenty-first-century global economy. These range from virtual currencies which operate within specific online communities such as users of a game or social network, to cryptocurrencies, which rely on cryptography to make their transactions secure. While there are an increasing number of competitors, Bitcoin was the first to produce a secure decentralized cryptocurrency in 2009. The way its founders achieved this was partly through the use of elliptic curve cryptography, but also by considering a particularly awkward mathematical puzzle called the Byzantine Generals problem.

This is the problem: there is a general with several commanders. In order to mount a successful attack on the castle they are besieging, the general has to give an order to attack or retreat via messengers. However, a few of the commanders and messengers are traitors who will give the wrong order. If the wrong orders are received, then the troops are liable to be defeated.

One way of getting around this problem is to require every commander to send messengers to all the other commanders, and no-one will act until each commander receives the same message from all the other commanders. Then the traitors cannot cause a false order to be relayed (see Figure 28).

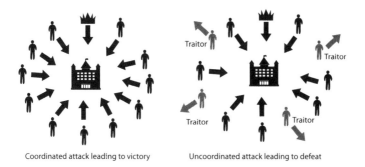

Figure 28. The general (represented by the crown) needs all his troops to attack the castle simultaneously. If traitors defect then the attack will fail.

Bitcoin needed to address a variation on this problem. For a self-governing distributed network to function and for everyone to trust in the currency without a third-party adjudicator, it is necessary that there is a consensus on the correct 'order' being given (which is here analogous to the record of all transactions and ownership of Bitcoin).

The solution to this problem is the root of the **Bitcoin blockchain**, which is a record of all transactions going back to the very start of the currency. This is basically a database which provides a consensus of replicated, shared and synchronized data spread over multiple sites. When you make a payment using Bitcoin your intended transaction is broadcast to every node of the network, which is made up of Bitcoin users and specifically Bitcoin miners. These are people who have set their machines to search for solutions to a computationally complex maths problem (in return for which service they are paid in newly created Bitcoins). Each node of the network examines the transaction to check it is legitimate, in other words that you are entitled to make it. At this point your transaction joins a queue of other verified transactions, which in turn forms the next potential block in the blockchain. Once a node solves the maths problem, this is combined with your transaction in a 'proof-of-work' and this is sent to the rest of the network for verification. If the solution is verified, the new block is added to the blockchain. It now contains a record of your transaction, along with the proof-of-work, and this is mathematically linked to the previous blocks in the chain.

Each new block adds cryptographically hashed data to the chain, and each is built upon the previous block, which ensures that the data in the blockchain cannot be compromised. This is because the proof-of-work would be so difficult to fake – in order to create a new blockchain which would fool enough of the nodes into accepting fraudulent transactions, you would need to reverse-engineer each link in the blockchain going back to the very beginning. Because this is clearly such an incredibly complex task, it is safe to trust that this process is producing an authentic record of legitimate transactions. (In theory it might be possible for someone who owned over 51% of the existing Bitcoin to manipulate the rest, which could have been a problem when the network was quite small; but once it reaches a certain size it seems implausible, if only because of the question of how someone would actually achieve this.)

There are other problems with cryptocurrencies. While the blockchain seems unhackable, individual wallets have been hacked, and the biggest hack was that on Mt Gox, a Bitcoin exchange based in Tokyo. Given that the value of the currency is partly based on people's trust levels in it, such instances are a concern and can cause crashes in its value. There are plenty of people who speculate on whether Bitcoin might be a kind of Ponzi or pyramid scheme in which original entrants have done well out of the process, but later entrants may end up losing money. It is also interesting to speculate on how international governments and corporations will deal with cryptocurrencies and the associated mathematics in future. Banking giants are slowly developing their own versions of the blockchain, while governments are concerned at the way that Bitcoin transactions fall outside of their jurisdiction – it seems possible that the fact that such transactions are being used for nefarious purposes could be used as an excuse to crack down on them.

But whatever the future of cryptocurrencies might be, it's hard to deny the brilliance of the mathematics that has gone into the creation of Bitcoin and other similar ventures.

Derivatives for Beginners

Peter Lynch's 6 out of 10 rule relies on the idea of a stock portfolio in which the most you can lose on a given investment is the money you invested. Of course the financial markets offer some far more complex investments than simple stock ownership. For a start, many brokerages allow margin trading, meaning they will advance you money to buy more shares than you have the immediate cash for. Margin trading also allows for **short-selling** in which you borrow shares and sell them in the hope that the price will fall and you can buy them back at a lower price in the future. Margin trading obviously comes with the danger of losing money you don't actually have, and it also allows for leveraged trading, in which the potential gains or losses are magnified.

Over recent decades there has also been a spectacular growth in the trading of **derivatives** such as futures and options (see below). These can be a bit harder to get your head around than straightforward share ownership. Essentially a derivative is a financial instrument whose value depends on a given asset – but ownership of the derivative does not give you ownership of the actual asset.

To get a sense of the basics of derivatives, imagine that Sam, the owner of Happy Farm, has a few problems he needs to solve. Firstly, his crop of apples is due to be ready for shipping in September. But the prices in recent years have been volatile and he needs to be sure he has enough money to buy the winter wheat for his livestock.

After negotiations, Sam's wholesale customer makes a promise in advance to pay £15 per crate of apples in six months' time, no matter what the market price is. This is how a **futures contract** works – whatever happens, Sam's winter wheat bill will still get paid. (This is an example of **hedging** in business – Sam has used the futures contract to hedge against volatility in the market.) If the market price increases beyond £15 a crate, the wholesaler will make the additional profit while Sam loses out on income he could have made. However, if the market is oversupplied and the price crashes to £5, then the wholesaler effectively takes the loss that Sam would have suffered.

The next problem Sam has is that Happy Farms needs another field and there is one for sale adjacent to their property. But because Happy Farms already has a large mortgage at a variable rate, the Common Sense Bank is worried about lending the business any more money. It would, however, be willing to make the loan if Sam could switch his debt to a fixed rate in order to ensure the payments stay at a constant level.

Sam approaches the owner of Pleasant Farm in the next valley, where they have a fixed rate loan on their property. They sign a deal for each to take over the other farm's loan. Now Sam can get the additional financing he needs, while Pleasant Farms are hoping that the interest rates will fall, improving their situation (but are willing to take the risk on rates rising for this potential gain). This is the underlying mechanism in **credit swaps**.

As it happens, Sam owns £50,000 of shares in Pleasant Farm and has a plan to sell these in a year's time. However, he wants to be protected against any fall in the price. So the Common Sense Bank agrees a deal where Sam pays them a fee for the right to sell them the shares in 12 months' time at their current valuation, again regardless of the market price at the time. If he chooses not to exercise this **option**, the bank has still made a profit on its fees. However, as an example of the risk that options can expose banks to, if the banks charged a 2% fee for this option (£1,000) and the market value of the shares fell to £30,000 the bank would now be £19,000 down on the deal (assuming Sam isn't silly enough to decline the option of selling for £50,000).

If this comes to pass, and the Common Sense Bank now has a sudden need to raise more finance, it may choose to sell ownership of Sam's loan on to a third party at a discounted rate. The bank gets its finance, while the third party gets the benefit of owning the loan at a better effective rate of interest. This is how **credit derivatives** work. For instance, if Sam's loan has an outstanding balance of £10,000 at 5% interest, and the loan is sold to InterCom bank for £5,000, then they are effectively collecting payments at a rate of 10%.

These are the fundamental types of trade that derivatives allow, although there are many varieties, including long and short versions of most trades. It should be fairly obvious that the main principle is that risk is being exchanged and spread out between those who need to be risk averse and those who are willing to take on more risk in pursuit of more profit.

The global financial crisis that started in 2007–8 has given derivatives a bad name, in some cases rightly so. Once debt is sold on repeatedly in different layers of derivatives, it can become increasingly unclear what is actually being bought and sold. Risk can thus be redistributed in ways that are dangerous to the entire system. Scott Adams, in a *Dilbert* cartoon strip, memorably compared some of the debt instruments that were being sold to investing all your money in diseased cows, on the basis that to invest in one diseased cow would be foolish, but aggregating them all together 'makes the risk go away'. This would be funny if it weren't for the fact that this is pretty much what was happening when high-risk loans were chopped up, repackaged and given high ratings.

There are parallels here with betting systems, such as the Martingale, which we looked at in Chapter 3. There we saw that, no matter how you slice and dice it, the original risk always stays in the system, and the sleight of hand that makes it seem to have disappeared tends to leave it lurking in a corner, with all of the losses waiting to be materialized given one particular outcome. This is what happened when the sub-prime market started to collapse in the US in 2007–8.

So derivatives do come with a health warning – Warren Buffett memorably described them as 'weapons of mass destruction' – but they also have a genuine role to play in redistributing risk and can be good investments if you truly understand the odds on both the up and down side of the trade.

Are Derivatives a Zero Sum Game?

If two people toss a coin, and the winner pays the loser £1, this is a **zero sum game.** The term comes from game theory where it indicates a game in which one party's loss is the other party's gain. It should be obvious that this applies to trades such as options and futures, if you ignore any fees that go to third parties. Some people suggest this means the entire derivatives market is a zero sum game in which risk is merely redistributed. Most individual trades in stocks can also be seen as zero sum games – the potential gain for the purchaser is a potential loss for the buyer, and vice versa. The reason we can't extend this logic to the entire stock market or the entire derivatives market is that wealth is created (and sometimes destroyed) by the activities of the entire market – the way that the stock market allows investment to flow around the economy and the way that derivatives allow the risk to be spread are both theoretically underpinning the process of wealth creation. And it is wealth creation that drives the long-term growth in the stock market as a whole. On the other hand we can apply the logic of zero sum games to the question of whether traders can 'beat the market'. For every trade that performs better than the market performance as a whole we should expect a balancing trade that underperformed the market as a whole, which is a salutary thought when it comes to assessing the honesty of those fund managers who claim to be able to consistently outperform the market.

The Black–Scholes Model and the Financial Crisis

The Black–Scholes model, developed by Fischer Black and Myron Scholes in their 1973 paper 'The Pricing of Options and Corporate Liabilities' is a way of analysing the value of derivative instruments in a financial market. This is the formula that led to a huge boom in options trading around the world in recent decades. Here is the actual equation:

$$C = SN(d_1) - (d_2)Ke^{-rt}$$

$$d_1 = \frac{\ln\left(\frac{S}{k}\right) + \left(r + \frac{s^2}{2}\right)t}{s\sqrt{t}}$$

$$d_2 = d_1 - s\sqrt{t}$$

where C is the call premium (the dollar amount over the par value of a callable debt security that is given to holders when the security is redeemed early by the issuer), S is the current stock price, t is the time until option exercise, K is the option striking price, r is the risk-free interest rate, N is the cumulative normal distribution, e is the exponential term, s is the standard deviation and ln is the natural logarithm.

It is somewhat daunting, so rather than go into every detail, here's an overview of what is going on. In 1900 the French mathematician Louis Bachelier proposed that it might be possible to model the zig-zag movements of stock prices as Brownian motion or a random walk. This is an example of the use of stochastic processes, which basically just means using probabilistic modelling of a random process over time. So we treat the price change at any moment as being random (but having a finite variance) – the mean of the movements over a particular period of time gives us the short-term average direction of price, whether up or down, while the standard deviation tells us how volatile the process is. The longer-term price movements tend, under these assumptions, to display a Gaussian (normal) distribution (see p. 41).

This leads to a fairly good model for how stock prices actually move, although it is inaccurate in extreme situations such as stock market crashes (in which any assumption of finite variance can be tested to destruction). Black, Scholes and Robert Merton started from this observation and worked up the equation to establish the price of an option. Note that the version above is specifically for European call options on non-dividend paying stock. These grant the holder the right but not the obligation to buy an asset at a particular price (known as the strike price) at a specified date.

The payoff for an option follows the hockey stick shape of graph (see Figure 29). This is because there is no payoff if the underlying asset doesn't rise to the strike price – thereafter the payoff rises with the price of the asset as we can sell it for an immediate profit.

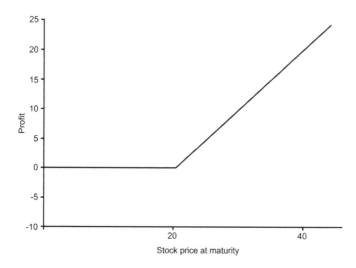

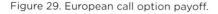

Figure 29. European call option payoff.

To calculate the value of an option we need to find a way to calculate expected value, for which we need the probability of the asset reaching particular prices. The first part of the equation, $SN(d_1)$, is an estimate of the value of the stock received, if any. Then we need to subtract the cost of the cash to buy the option. This is the second part, $KN(d_2)e^{-rt}$: the e^{-rt} part of this is a discount factor to account for the time value of money (bearing in mind that money is worth less to us in the future than it is now). Finally, the equations for d_1 and d_2 look quite complex but are essentially there to allow for the volatility of the asset. An option for a highly volatile asset is worth more than that for a less volatile one – this makes intuitive sense because a large swing down in value won't cost us any more than a small one (if both take us below the strike price), whereas a large gain in price is more valuable to us than a small gain.

The Black–Scholes equation makes a number of assumptions, notably that volatility is a constant, there are no transaction costs or limits on short-selling and that we have a reliable value for the risk-free interest rate. These are all inaccurate in some respects but the equation nonetheless does a fairly good job in normal conditions at valuing options.

Quants and analysts have developed a wide variety of similar equations for valuing other derivatives, and this has underpinned a huge increase in derivatives trading, simply because the equations provide a way to have some objective sense of the value of those derivatives. Some have explicitly blamed the Black–Scholes equation for the global financial crisis, because the equations provided by the quants gave traders a false confidence – it's certainly fair to note that the crisis demonstrated some of the dangers of the derivatives market and the limitations of mathematical analysis of the markets. I've already mentioned the phrase 'put garbage in, you'll get garbage out'. The trouble with derivatives equations is that the people using them can forget that they are only a model and that in certain situations the actual equation might turn out to be garbage, not to mention the data that you are inputting and outputting. Any mathematical financial model is at best an estimate, and there should always be a plan B for situations in which the modelling fails.

The Maths of High-Frequency Trading

Since Ed Thorp started Princeton Newport Partners in 1969, mathematicians have gone from being a boring but necessary part of the risk management and accounting departments to having a central role in the global finance system. Quants play a central role in our financial capitals, through algorithms such as the Black–Scholes model, high-frequency trading and a variety of other applications.

The rise of mathematics can be illustrated by a look at the finance career of a maths genius called Jim Simons. Within the academic sphere, Simons is known as the co-creator of the Chern-Simons 3-form, an advanced piece of topology. But he is also known as the founder of the wildly successful hedge fund management company, Renaissance Technologies. In October 2015, Renaissance controlled $65 billion worth of assets, most of which belonged to its employees.

The company recruits many of its employees from academia and theoretical research rather than from the world of finance – a third of them have PhDs in subjects such as mathematics and physics. For instance, since Simons retired, the company has been run by two computer scientists specializing in computational linguistics who joined the company from IBM Research. Its investment strategies are heavily dependent on complex algorithms and mathematical models of the world. And this rarefied approach has made it one of the most consistently successful hedge funds around.

Quants have also been a crucial part of the story of **high-frequency trading** (HFT). This is an area of finance that has only been possible since it became possible to make digital trades. The companies using this strategy make extremely short-term trades. In *Flash Boys*, the author Michael Lewis talks of their battle to get the fastest possible connections in order to gain the crucial advantage of a few milliseconds over their rivals. Some are even using artificial intelligence programs that can learn from experience and adapt their strategies as a result. The actual mathematical strategies being used tend to be the same as those used by any short-term speculator – hedging, arbitrage and taking advantage of short-term momentum shifts – but rather than doing this over hours, days or weeks, they are doing it in milliseconds. And this also means that they can sometimes respond to a situation by seeing that others are about to make a purchase and getting in there first (which in turn feeds the dangerous feedback loop of 'spoofing' that is designed to lure the HFT types into rash moves).

There is much debate over whether such trading increases volatility in the market or brings a stabilizing factor. There is an obvious danger in thousands of computers carrying out automated trades. The Flash Crash of 6 May 2010 was largely caused by the self-reinforcing feedback loops of such automated trading. While some measures have been introduced to serve as a brake on such events, some still regard HFT as a dangerous practice. Charlie Munger, for instance, has argued that it is 'very stupid to allow a system to evolve where half of the trading is a bunch of short-term people trying to get information one millionth of a nanosecond ahead of somebody else'. On the other hand one UK study found that quant trading helped reduce dealing costs as well as improving liquidity and concluded that it did no harm to the efficiency of the market.

Like it or not, HFT is probably here to stay. It is also an example of how rapidly the world of finance can change – as a strategy it was virtually unknown in the 1990s, but it grew rapidly to a large part of the market, with the early participants making huge profits. Then, as more and more funds moved into the area, it became less lucrative and profits have fallen. This is fairly typical of the digital economy, and a similar process has been seen in many other sectors in which maths is crucial.

So if you have high-level maths ability, while becoming a quant or a HFT trader may still be a path to wealth, it might be better to recognize that the biggest gains go to the innovators and to start looking now for the 'next

Google', the 'next HFT', 'the next Bitcoin' or the next breakout technology.

Chapter 6 Summary

1. Algorithms and quants rule the world. Who will rule the world in the next few decades? We probably don't know yet but the chances are they will be relying on maths in one way or another.
2. The relationship between maths and finance is getting ever more complex.
3. Some of the future moneymaking opportunities (and instabilities) in the global economy will be directly related to this complexity.
4. The maths involved in digital breakthroughs isn't always PhD level (although some of it is...).

CHAPTER 7

Use Maths to Improve Your Performance

One should study mathematics simply
because it helps to arrange one's ideas.
M. W. Lomonossow

A big advantage of mathematical thinking is that it is rigorous – one way we can employ maths positively is to use it to make better decisions and to better understand patterns. But we can also use mathematical thinking defensively, to guard against irrational thinking and poor decisions by relying on the maths and by using correct statistical approaches.

Gather Sufficient Data

Too many bad decisions happen because people rely on insufficient data. A proper understanding of statistical significance and sample size can lead us to gather sufficient information to make a good decision, and to seek out data sources other than those which we know will confirm our biases. At the very least this can tell us when a business or investment decision is being made for mathematically sound reasons, rather than for more emotional or instinctive ones.

It's always worth bearing in mind the laws of small and large numbers (see Chapter 3), which suggest that we are far too ready to jump to conclusions from small samples. If, for instance, you are looking at sales figures for a single area over several months, it is fairly unlikely that you have sufficient data to judge whether the overall trend is upwards or downwards. And bear in mind that standard deviation will tend to produce local variation in the figures which doesn't reflect any particular cause.

When it comes to scientific experiments, statisticians use the concept of **statistical significance**. Given two variables, a result has statistical significance if it is very unlikely to have occurred given the null hypothesis (which is the hypothesis that the two variables aren't correlated). To be more specific, we choose a significance level, α, for each experiment – this is the probability of the study showing a correlation where there isn't one. We then compare this to the p-value of a result, p, which is the probability of such a statistical result being shown by chance. The results are said to be statistically significant if the p-value is lower than the chosen significance level. We often use 5% as a reasonable measure of statistical significance – we can also subtract the significance level from 100% to give a confidence level – so a 5% level of statistical significance means we have a 95% confidence level.

It can of course be hard to apply true scientific rigour to sales figures and market research, but we can bear in mind the underlying principle here – which is that the less likely it is for a set of figures to have occurred by pure chance, the more likely it is that we can identify the underlying reasons with confidence.

The next thing to remember is the old cliché that correlation doesn't imply causation. If we have a high degree of confidence in a set of figures, we still need to consider all the possible explanations. For instance, if one sales rep's figures have increased across all of their territories, our first instinct might be that the rep is doing a great job. However, if the company's big new product is a seed planter and the rep works exclusively in rural areas, we should probably conclude that the sales figures are caused by the new product launch rather than the excellence of the rep.

The main things to consider when examining a correlated set of figures are whether the sample is big enough, whether the connection is sufficiently strong and consistent, and whether there are any other possible explanations. For instance, you would probably find a correlation between the average amount of time that male and female employees take for their lunch break. However, one doesn't cause the other – both are caused by the separate factor of the company rulebook.

When it comes to marketing, it will often pay dividends to think in terms of experiments and hypotheses. If, for instance, you have a social media presence and want to look at new ways to drum up business, then try starting with a concrete theory about what will succeed in achieving this aim, for instance a change to the message on your profile. Then compare the results of making that change for a set period. Don't think only in terms of A/B tests in which you test two ways of working against each other: try instead to think of everything that might possibly be statistically significant and measure it if you have the opportunity to make a testable change in your methods. If you're testing ideas to increase conversion from clicks to purchases, then keep records of all the data that might be relevant to that conversion rate – for instance, what was the tone of the 'clickbait'? How explicitly was it selling the product? What kinds of picture did it rely on?

Also keep checking that the results of any change you have made are consistent over subsequent sales cycles. It's not enough that a change coincided with an improvement in the click-through rate – you need also to confirm that the effect isn't a temporary one that might have been caused by other factors.

At the end of the day, you may not be able to take a completely scientific approach, but you will have one that is far more robust if you think in terms of hypotheses and statistical significance from the start.

What to Do

When it comes to figures, make sure you aren't reaching conclusions from an inadequate sample size. Gather as much data as possible – you may find it useful to get training in your company's accounts system or sales database, as you can often generate your own reports from these. Remember that any future projections are only useful if they come with a high confidence level. And consider all possible explanations for any data.

Understand Randomness

As we have seen in earlier chapters, humans are bad at allowing for randomness because we instinctively look for patterns. We are also not good at emulating random behaviour. The classic example of this is that we wrongly assume that a long run of heads in a series of coin tosses makes a tail more probable on the next throw. Avoiding misunderstandings of random distribution can help us to avoid common errors, like assuming that correlation automatically equates to causation, even when the variation in the figures is not statistically significant.

Fooled By Randomness by Nassim Nicholas Taleb is an interesting meditation on this problem. His essential argument is that randomness, chance and luck affect our lives in many ways that we fail to understand. Hindsight bias and survivorship bias mean that we assume that individuals always become successful through skill rather than chance – he contrasts a wealthy dentist, who would have become successful in most alternative versions of reality, because he has done the arduous training in a field that is generally rewarding, with the winners in more volatile areas such as finance, where the wealthy fund manager might have been wiped by a fairly small change in the course of events. His point is that the quality of our decisions should be partly judged by those alternative scenarios, not just by the actual outcome. For instance, someone who makes a wrong call that the stock market has reached a high may still have a correct analysis of the fundamentals, while someone who calls the market correctly may have been lucky. The most reliable and certain kind of success is exhibited by someone who would succeed in the greatest proportion of different alternative histories.

There are a lot of lessons here for anyone in business or investment. The short-term information we use to make decisions is often just statistical noise. So when it comes to assessing a strategy that has been productive, it makes sense to repeat the strategy, but also to make sure we aren't thereby exposing ourselves to a high downside risk, since it takes a long period of repetition to be sure that it is the strategy itself that is causing the success, rather than random variation. Then, over time, the repetition will show us whether we are actually making the right call, or whether it was mere luck.

Taleb also emphasizes that it can be worth risking small losses for big gains. Since highly unlikely events can be underpriced (as they were for the Hole-in-One gang, see p. 185) it is sometimes worth a small gamble on something which has a high potential reward. As Taleb notes, the expected value from an option which loses us money in 95% of scenarios, but makes a huge profit on the other 5%, may still be significantly positive. And this can be a less dangerous way to trade than to take a risk that will make a small amount of money 95% of the time but leave you at risk of ruin in the other 5% of possible outcomes.

We seem to be programmed to misunderstand randomness, so proper understanding of its workings can only be gained the hard way – through repeated consideration of statistical concepts such as standard deviation and the law of large numbers. For instance our instinct seems to be to see a series of coin tosses of tails/heads/tails/heads/tails/heads as being more random than tails/tails/tails/heads/heads/heads – we perceive the first series as more authentically random simply because it switches sides more often. This is why Taleb's advice to always bear in mind that something may be the result of randomness is a useful discipline – time will eventually show us whether an effect is random or the result of our strategies and decisions, but in the short term it's best to assume that either might be the crucial factor.

Use Visualizations (But With Care)

There was a time when many mathematicians were sceptical of visualizations. When, in the late nineteenth century, Giuseppe Peano described the space-filling curve, an infinite fractal which can be drawn within a square, he used equations rather than a simple series of pictures. It took later mathematicians such as David Hilbert to demonstrate the beauty of the shape he had been describing.

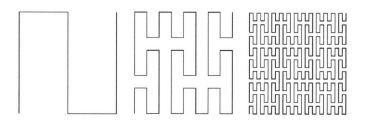

Figure 30. The first three iterations of Hilbert's version of the space-filling curve. Theoretically, as it becomes infinitely complex, the single line touches every single point of the square.

However, mathematics has become gradually more comfortable with visualizations as the range of subjects studied has expanded. The Iranian maths genius Maryam Mirzakhani, who sadly died in 2017, is an example of a mathematician who worked largely visually, using diagrams and patterns throughout her work. She was a talented artist and, within mathematics, studied different forms of geometry, abstract shapes and structures in higher-dimensional spaces as well as more prosaic problems such as the path a billiard ball will take around a polygonal table (which is a surprisingly complex maths problem).

In business and investment you are unlikely to need to be able to visualize a hyperbolic, multidimensional billiard table, but visualizations are nonetheless a hugely important way of processing information. A chart can be more persuasive and easier to understand than a spreadsheet. A flowchart showing the production line or critical path for a new product can help to clarify the process. Visualizations can be the most powerful way both of understanding patterns in data and of presenting these patterns to others. But beware of distortions – for instance graphs which don't have zero for a baseline and thus exaggerate changes, or Venn diagrams that misrepresent the size of different groups.

In this section I'll give examples of some common errors introduced by mathematical visualizations, which can be almost as dangerous as failing to consider the maths in the first place. Understanding these distortions can help you to see the faults in others' arguments, and to present your own with greater clarity.

For instance, look at the chart of the heights of giraffes (Figure 10). When you first read that chapter, you should hopefully have noticed that this chart is mathematically inaccurate. The baseline isn't set at zero – if it was, the giraffes would be much closer to each other in terms of the ratio of their heights. The smallest giraffe on the chart looks to be less than half the size of the tallest, but the actual range of heights is 153–177cm. It's drawn that way because it seemed funnier and more dramatic, but if we had wanted a true picture of the difference between the giraffes, we would have had to draw it again with the baseline at zero.

The non-zero baseline is a common example of how visual representations of data can be misleading. It's fairly harmless in the case of the giraffes, but imagine a sales director presenting Figure 31 to a management meeting. It looks as though sales have doubled over the period, where in fact they have only increased by 12.5%. So this would be an inaccurate and highly flattering way of presenting the figures.

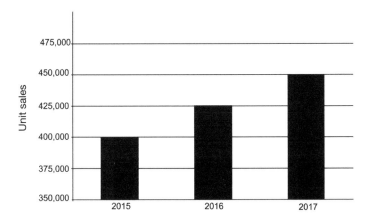

Figure 31. Sales graph.

If the sales director wants to make it look even more dramatic, he might use three-dimensional bars as in Figure 32, which exaggerate the rise to an even more extreme degree. These kinds of charts are always worth a touch of scepticism, because the three-dimensional effect distorts the actual height of the columns and makes it hard to read the heights of the bars in relation to the scale.

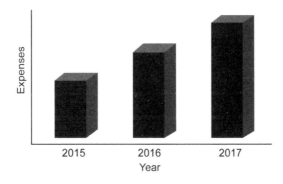

Figure 32. Sales graph with three-dimensional bars.

It's always worth checking the context in which any figures are being presented. Figure 33 (overleaf) is a chart provided by the production department of Plastic Bananas Ltd, giving the percentages of customers who gave different reasons for products being returned.

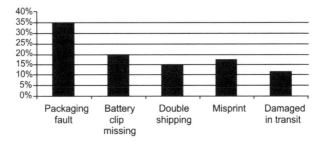

Figure 33. Chart of reasons for returning packages to Plastic Bananas Ltd.

On its own this is a fairly useless representation: it could for instance imply that 35% of **all** products are being returned because of packaging faults. To make any sense of it, we would also need to know what percentage of products are being returned in the first place. If this figure is running at 0.1%, then just 0.035% are being returned with faulty packaging, which would make it a far less significant concern.

It's also important to take into account whether the data being presented is sufficiently complete. If that sneaky sales director wants to big up their performance, they might present the chart shown in Figure 34.

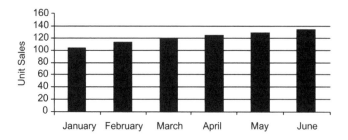

Figure 34. Sales figures, January to June.

The upward trend in these figures is undeniable. However, compare it with the graph of the twelve-month period shown in Figure 35.

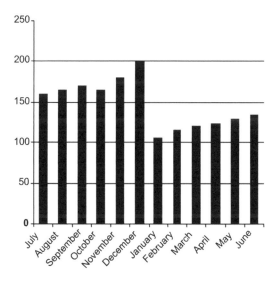

Figure 35. Sales figures, twelve months July to June.

Now the situation looks a lot less healthy – the increase since January could simply be the transition from the quietest part of the retail year to the summer, and the current figures for June look bad in comparison to the rest of the summer months last year. To get a sense of what is really going on, we'd ideally also need to see the full figures for several previous twelve-month periods, in order to make a clear comparison. As ever, the more data we have, the clearer the picture will be.

Figure 36 shows another chart where the axes are a problem, but in this example it is because there are two y-axes. In this case the sales director has been challenged on the rise in expenses claimed by the sales reps. In response, he claims that there is a relationship between

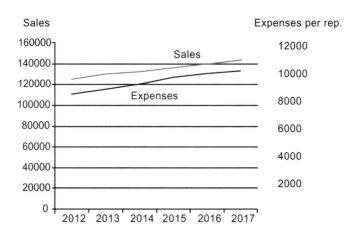

Figure 36. Graph showing sales and expenses.

expenses and sales – which the chart appears to bear out. However, the use of two y-axes, with different scales, is an obfuscation – there has been only about 8% growth in sales (the upper line), but approximately 25% increase in the expenses claimed per rep (the lower line). It would also probably make more sense to present the figures as a total expense rather than per rep.

In the next instance, a Venn diagram has been badly constructed (Figure 37). The aim is to show how often the purchasers of toys in a particular store also buy clothes during the same visit. However, the figures are unclearly defined (10% and 20% of what exactly?) and the sizes of the circles within the whole diagram are clearly misleading. Based on area, it would seem that less than half of toy purchasers also buy clothes, but the comparison in the figures suggests that there is a crossover in 80% of toy purchases. As with the previous charts, an accurately presented version of this image could be a useful aid to comprehension, but the flawed version is simply confusing.

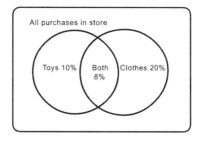

Figure 37. Venn diagram showing information on store purchases.

What to Do

Do use visualizations, and learn how to make them accurate. Also learn to spot errors and obfuscation in the visualizations that colleagues use (especially that pesky sales director)...

Listen to the Data

Facts can be boring. For instance, the long-term maths of owning and renting property or index funds is not a particularly gripping story, but those who listen to the relentless long-term trend of increasing prices in such markets have often been successful: consider the story of the Candy brothers, property investors who started with a £6,000 loan for a London property in 1995 and ended up fourteen years later on the *Sunday Times* Rich List with a portfolio worth £330 million. Of course luck and timing also play a crucial role – 1995 was a point when prices were still below their long-term average, and they went on to rise to record highs. Buying property in Japan during the 1986–91 bubble or in Ireland before the property crash there in recent years would have led to a far more disastrous outcome.

Many of Warren Buffett's most successful investments are boringly obvious, because he does laborious research, looks at the maths and the facts, and trusts in them. The same thing is often true in business. The best strategy for a company may be a very boring one – to simply keep doing the core things that they do well. Having a mathematical understanding of how the results of that business have changed over time can help to identify whether it is viable in the long term or needs an overhaul. And having an understanding of randomness and risk can help to identify whether a particular new direction is a good bet or a bad bet.

Another person who listened to the facts with spectacular results was Andy Krieger, a currency trader at Bankers Trust. In 1987, at the age of 32, he took a short position against the New Zealand dollar that earned him millions – he did this by trusting in the data in spite of misgivings. At the time a range of currencies were rising against the dollar following the Black Monday crash. A huge amount of capital fled from the American dollar into apparently safer currencies. In spite of some initial doubts, Krieger trusted his analysis of the data, which suggested to him that this would lead to those other currencies becoming overvalued, creating an opportunity for arbitrage.

Using options, he took out a short position of hundreds of millions of dollars against the currency (also known as the kiwi). The kiwi ended up falling back by up to 5%, giving Krieger's employers million-dollar profits. Krieger went on to work for George Soros who also made huge profits from betting against currencies, although he has since expressed regret for the damage that can be caused this way (since it can be damaging for the economies of smaller countries to be subjected to this kind of currency raid).

Similarly, Michael Lewis's book *The Big Short* shows how a few financiers listened to the data about mortgage-backed securities and the state of the US housing market in the run up to the global financial crisis. One such financier, Michael Burry, manager of the Scion Capital hedge fund, examined the sub-prime market in 2005. He carried out extensive analysis of data about the mortgage lending practices of the previous three years, and correctly concluded that this was a bubble waiting to burst. He persuaded Goldman Sachs to sell him credit default swaps against sub-prime deals. At one point he faced down a revolt from his investors, who were nervous that he had made a huge mistake. In the end, the sub-prime market tanked, earning Burry profits of $100 million as well as $700 million profit for those untrusting investors.

Of course, a delay of a few more years in the market crash could have left Burry with a serious headache, so you need a high degree of confidence in your data to take such huge risks. And if we wanted cautionary tales we could tell the story of plenty of traders and investors who took large risks and lost millions. But the real moral here is that, while you may not be able to find ways to make $100 million, you can certainly prosper in more realistic ways on an everyday basis if you simply observe the mathematical data and trust in what it is telling you.

Apart from anything else, it's much more likely to be telling you the truth than some of your colleagues are.

How to Sell a Ticket for the St Petersburg Lottery

First impressions are hugely important in business. If you present your boss and colleagues with a spreadsheet of deals in which all the least profitable deals are at the top, they are likely to be less impressed than if you order it with the most profitable deals to the fore.

In Chapter 3 we looked at the St Petersburg lottery, in which we ask how much a buyer should pay for a ticket where the prize is £1 if the first time a coin toss lands on heads is the first throw, £2 if it's the second throw, £4 if it's the third throw, £8 if it's the fourth and so on, with the prize doubling each time. At college I ran a psychology experiment based on this problem. A control group were presented with the problem as above. A second group were given a different presentation. They were told that the prize would be £500 if the first coin to land on heads was the tenth coin, with the prize doubling thereafter. Then they were told that there were also consolation prizes ranging from £250 down to £1 for any games that didn't reach the tenth coin.

The second group turned out to be willing to pay an average of 60% more than the control group, even though the terms on offer were slightly worse. The conclusion was that, when it comes to figures, inviting the buyer to focus on the larger prizes first was a more effective psychological strategy than doing the opposite. Which is why people buy lottery tickets in the first place – because the idea of winning a million is much more powerful than the knowledge that this almost certainly won't happen. No-one advertises a lottery by showing someone who is mildly pleased to have won £2 on a £1 scratchcard.

In other cases people make losing bets for more rational reasons. Insurance, like many other businesses, involves selling something for more than it costs you. In the case of the insurer, they receive insurance premiums, which are set at a level that hopefully guarantees they will pay out less than they receive. If they couldn't do this it wouldn't be a profitable business. So why does anyone buy insurance, knowing this to be the case?

The thing is, when you buy home insurance or car insurance, it doesn't matter that you are effectively paying out more than you are likely to get in return. The utility you gain from knowing that you won't be ruined if the worst comes to the worst is greater than the cost. In theory, for a billionaire who could afford to take the costs of a house fire or stolen car, it would make sense to go without insurance as, over a lifetime, they would be likely to be better off without it. For the rest of us, insurance is a bad bet, but a bad bet that is worth making.

Trade Fair Maths

In a previous career, I used to travel to trade fairs at which we had a large number of short meetings with customers and suppliers. Early on my boss advised me to use 'trade fair maths' to come up with an estimated price on one of our products. Initially I didn't have a clue what he was talking about, so during a break between meetings he explained that one of the most useful things a salesperson could have was a good grasp of how to use mental arithmetic to come up with estimates.

For instance, we needed to make about 30% margin on one of our products, which normally cost £1.75 to produce. (To calculate margin m, you subtract production cost, p, from the price received [selling price, s] and then express the result as a percentage of price received, that is, $\frac{100(s - p)}{s} = m$, which can be manipulated to $s = \frac{100p}{(100 - m)}$. So to accurately add 30% margin to production costs of £1.75, for instance, divide by 7 and multiply by 10 giving £2.50.)

If the customer asked for a different specification to the one I had costed, I might need to adjust our cost price up by 10% then give them a ballpark figure for what they would be paying. It's easy to get flustered doing this kind of calculation on the spot, but I got quicker and more confident once I started to use a few shortcuts. To add 10%, I adjusted upwards in my head to £2.00, then multiplied by $\frac{3}{2}$ to get £3.00. The actual figure is £2.75, but I would at least be able to give them a quote on the spot (and promise to follow up later with a refined quote).

Note that both of my shortcuts padded the price slightly upwards. In trade fair maths it's important not to use estimates that pad your figure in the wrong direction. If, for instance, I had multiplied by $\frac{4}{3}$ I'd have ended up with a figure of £2.66 and would have been making too cheap an offer, which I would then have had to disappoint them by withdrawing. Better to quote a bit too high and then negotiate downwards rather than vice versa. Similarly if I had been making an offer to buy a product I would have needed to pad any estimates downward to give myself room for error.

There is a trick to this, which is to know the ratio you should be applying and to quickly check that the shortcut you use is slightly more generous, on the right side. For instance in this case the correct ratio would be to multiply by $\frac{10}{7}$. Multiplying by $\frac{3}{2}$ will give a slight overestimate because, as a moment's thought will show, $\frac{3}{2}$ is equivalent to $\frac{10.5}{7}$. Whereas when we compare $\frac{4}{3}$ with $\frac{10}{7}$, we can convert the denominator to the lowest common multiple, 21, and see that we are comparing $\frac{28}{21}$ with $\frac{30}{21}$, and that multiplying by $\frac{4}{3}$ will be an underestimate.

Padding is a useful principle in a variety of business situations, if applied carefully. For instance at the same business we manufactured products in Europe, China and America and sold around the world. On our cost projections we would add 5% to any production costs that had been quoted in foreign currency and subtract 5% from our projected income for any invoices that were to be paid in foreign currency. This protected our margin from currency variation and, if the currencies stayed stable, gave us a nice surprise when we compared the final results to the projections. Of course, if you pad too much you can persuade yourself out of deals that might otherwise be viable, or end up quoting too high and losing business, so it's a strategy that needs a bit of common sense and flexibility.

In a similar vein, when your plumber or car mechanic quotes you for a job that requires parts they will often be padding the actual costs – in which case it makes sense to haggle and try to establish the exact price of any parts that are to be added on to the labour.

Padding can also be a smart move in investment. Firstly it helps to protect you from overoptimistic projections. Secondly, one of the costs investors face is the commission and fees that brokers take from their trades. While there are more and more affordable ways to buy and sell assets, this can still be a significant cost. Finally the mechanics of stop orders and market orders mean that you may not always get the exact price that appears to be available, since these orders will sometimes be filled with purchases that don't quite meet your target price. Again, having padded your projections in the first place can help to protect you from a negative outcome.

What to Do

Practise your mental arithmetic, especially when it comes to making a reasonable estimate of a complex calculation. If you are in high-pressure work situations you may be able to use this to your advantage by producing quick ballpark figures. It can also be useful when presented with spreadsheets and costings in meetings – it never does any harm to be able to quickly calculate the total cost of a project under discussion.

Playing the Game

In Chapter 3 we briefly encountered **game theory** when we were looking at the Keynesian beauty contest. The eccentric Hungarian mathematician John von Neumann is remembered as the father of this branch of mathematics.[*] It was initially inspired by his fascination with poker and the way that people make decisions when they have incomplete information. Whatever line of business you are in, it's worth studying some basic game theory as it has some interesting lessons about how negotiations work.

In game theory, the alternative decisions available to each player in a game are shown in a matrix. For

---

[*] To be pedantic, Emile Borel, a French mathematician, had actually published earlier work on the subject but it was von Neumann's work that established the theoretical framework.

instance, imagine Bob has to agree a price to buy a house from Jenny. The payoffs they can each expect from two alternative negotiation strategies for each of them are shown in Figure 38.

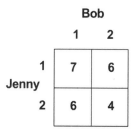

Figure 38. The price paid for a house depends on the strategies adopted by Bob and Jenny.

For many simple games we can apply the **minimax theorem**, which states that you should pick the strategy where the maximum advantage of your opponent is minimized. In this case Bob will pick strategy 2 to avoid having to pay 7, while Jenny will pick 1 to avoid receiving only 4. When we look at the square in which these strategies coincide, we find that the solution to the game is that Bob pays Jenny 6.

Similar tables of strategies can be used to determine the best way to negotiate an agreement, and it is often important to take into account which order the players make their choices in. For instance look at the table of strategies in Figure 39. In each case the figures represent the payments the players will receive (so '3, 4' in the top left box represents a situation in which Carol receives 3 while Alice receives 4).

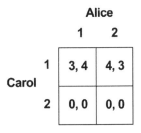

Figure 39. The order of play may determine the outcome of a negotiation.

It seems that Alice, playing first, can choose strategy 1 to get the best outcome. However if Carol finds a way of making the threat that, should Alice choose strategy 1, she will choose strategy 2 (and if Alice finds this threat credible), then Carol can force Alice to choose strategy 2, leading to a win of 4 instead of 3. On the other hand, if Carol plays first, she can only be sure of winning 3.

Next, consider a different kind of negotiation (see Figure 40). Tim, playing second, can hope to achieve the best total outcome by choosing strategy 2 – but this is only achievable if he can persuade Betty to choose 2 as well. So it is now in Tim's interest to make a promise that he won't be greedy and pick strategy 1 if Betty makes the first move. Otherwise she will pick strategy 1 to avoid a loss, and neither party will make a profit.

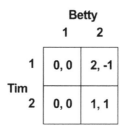

Figure 40. Can Tim persuade Becky to pick strategy 2?

Obviously most real-world negotiations are far more complex than those represented here. But an understanding of game theory can still be useful when it comes to understanding how you can force the best outcome even if you have limited choices. It's also important to note that the way the negotiation is conducted (and who makes the 'first move', such as a starting offer) can have a huge impact on the outcome.

The Mathematics of Wealth Distribution

Earlier in the book we saw a few of the reasons why it is easier for a rich person to make money than for a poor person. In the Rich Man vs Poor Man game in the casino, we saw how much more likely the poor man was to lose their money. And given the way that assets such as property and stocks and shares have increased in value over decades, it should be fairly obvious that those with a large amount of wealth to invest are in the best position to increase their wealth significantly, without having to do anything much more complicated than keep their wealth in long-term assets. So the most honest advice if anyone asks about the best way to make a million is 'start out rich'.*

There are a few different mathematical models that are used to measure wealth distribution. The standard measure of wealth inequality in a country is the Gini Index. Imagine a country with ten people whose income adds up to £100,000, and five ways that income could be distributed (see Table 7). The letters on the rows represent different people (or groups of people) and the numbers on the columns represent different distribution patterns.

	1	2	3	4	5
A	10,000	5,000	3,000	2,000	0
B	10,000	6,000	3,000	2,000	0
C	10,000	7,000	3,000	3,000	0
D	10,000	8,000	4,000	3,000	0
E	10,000	9,000	5,000	3,000	0
F	10,000	10,000	7,000	5,000	0
G	10,000	12,000	10,000	6,000	0
H	10,000	12,000	14,000	12,000	0
I	10,000	14,000	21,000	24,000	0
J	10,000	17,000	30,000	40,000	100,000

Table 7. Five different ways that income come be apportioned between ten individuals (A to J).

* There are a few variations on this advice – a common joke in film production companies is that the answer to 'how do you make a small fortune in the movies?' is 'start out with a large fortune'.

If we order the income earners from lowest to highest (as in Table 7) and plot the cumulative incomes on a graph, we get the graph shown in Figure 41.

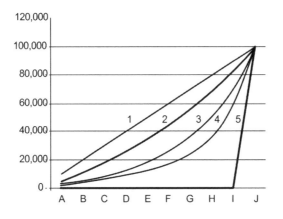

Figure 41. Wealth distribution graph: cumulative incomes of people A–J for each of the five distribution patterns.

The Gini Index treats the area between line number 1 (representing absolutely equal incomes) and line number 5 (representing the entirely unequal distribution of the final set of figures) as 100%. Then for each of the other curves it measures what percentage of this area lies between the curve and the straight line on the left – the resulting figure is used as a measure of inequality: the closer the index is to 100% the more unequal the society. Of course it's worth bearing in mind that this measure on its own doesn't tell you anything about how wealthy a society is – indeed there is a good argument that poorer countries, with very few rich people, will inevitably have a more equal wealth distribution. But the Gini Index can still be useful when measuring changes in a particular country over time, or comparing similar countries.

Another way of analysing wealth is the Pareto Principle, named after the Italian economist Vilfredo Pareto, who had noticed that 80% of the peas in his vegetable patch came from 20% of the pods. He went on to show that this 80/20 distribution was quite a common one in many other situations and in particular it was a good yardstick for how wealth tended to be distributed in society. In general 80% of the wealth tends to belong to 20% of the population.

More generally the Pareto Principle is the observation that many things in life are distributed unequally, and in particular that 20% of the input tends to be the cause of 80% of the result. This idea can be applied in a variety of ways. Firstly, it suggests that 20% of employees will be responsible for 80% of productivity, and that 20% of your customers will provide 80% of the income (which has some obvious consequences for how you manage a business). In IT firms, it has been observed that if you fix the worst 20% of the bugs in a system this will help you to avoid 80% of the crashes.

When it comes to the workplace, it has an application to time management. We spread our efforts out thinly over a range of projects even though the 20% that are the most promising will lead to 80% of the final success.

(Bear in mind that this is only a rule of thumb, and is not called the 80/20 rule because 80 and 20 add up to

100. The two numbers measure different things, so it can equally be the case that, for instance, 80% of the outputs from a system are the result of 10% or 30% of the inputs, or that 75% are the result of 20%.)

A really fascinating thing about the Pareto Principle is that when you plot these patterns on a graph, they follow a power law distribution. This means that they behave like fractals, in that any part of the distribution shows self-similarity to the whole distribution. For instance, if 20% of the population own 80% of the wealth, you will also tend to find that 20% of the wealthiest 20% own 80% of their portion of the wealth. In other words 4% will own 64% of the wealth. And 0.8% will own 51.2%, and so on. Note from Figure 42 that this implies that the top 20% of the bottom 80% control a similar amount of wealth (16% of the total) to the bottom 80% of the top 20%. (Actually there will be a slight slope between any two adjacent segments of the curve, so this can only work as a rule of thumb – it's more likely for instance that the richest 20% of the poorest 80% will earn 15% of GDP while the poorest 80% of the top 20% will earn 17%.)

Another notable aspect of power law distributions is that they often have a 'long tail'. In the short term this means that a company manufacturing many products will find that a small proportion of them are bestsellers, but many have increasingly small sales. However, seen from a broader perspective, the long tail can be a positive – a product with a long tail is one that sells well initially, then sales reduce but it goes on selling for a long time. Companies always need new products with high sales, but these are expensive to develop and it is often the earlier products with the long tails that keep on paying the bills in the meantime.

So while the 80/20 rule can seem depressing on first sight, as it suggests that life simply isn't fair, it does have some useful lessons about how to think about a number of aspects of our lives and work.

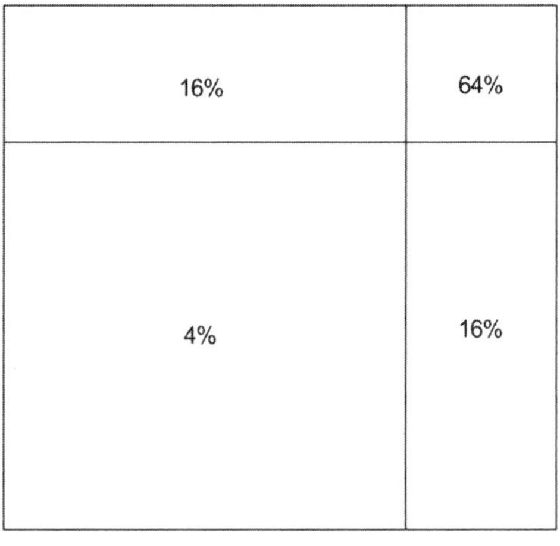

Figure 42. The size of the area in this image indicates the proportion of the population – the bottom 64% have 4% of the wealth. The next two tranches of 16% of the population (top left and bottom right) hold approximately 16% of the wealth each, while the top 4% hold 64% of the wealth.

What to Do

When it comes to making money, the most crucial thing to take from the Pareto Principle is that, no matter where you are on the wealth distribution graph, and no matter how you are trying to earn a living, it is always worth asking whether you can devote a greater proportion of your time to the most productive parts of your activities. However, be aware that it isn't always easy to identify the most useful activities in advance. I was once in a board meeting where a non-executive director (who was perfectly nice but perhaps not the sharpest tool in the box) asked a well-meaning question. He observed that most of our profit came from a small percentage of the products. So why, he asked, did we bother making all the other products? Why didn't we just make the bestsellers? Of course the answer was that no-one knows which products are going to take off until you've already made them...

Debt and Leverage

If you don't happen to already be a member of the super-rich, then at some point in your life you may need to use debt. If, for instance, you buy a car with a loan or a house with a mortgage, or shares on a margin account, you are using **leverage**, which is also a crucial tool for hedge funds and other financial institutions. Essentially, leverage is a way of referring to how much of your investment has been made using borrowed money.

It's important to understand how leverage increases both your risk and potential reward. If you make a £20,000 downpayment on a £200,000 house, you have paid 10% of the purchase price, so the purchase is leveraged 10 : 1. (The derivation, from 'lever' is a reference to the way that levers can be used to lift a far greater weight than you could lift without them.)

The upside of leverage is that it allows you to make purchases you would otherwise be unable to afford and magnifies any gains. If the price of that house now increases by 10% to £220,000, you have made 100% profit on your actual investment. The downside is that it can equally magnify losses. If the house declines in value by 15% to £170,000, then you have lost your initial £20,000 plus the house is worth £10,000 less than your outstanding debt. This is the reason why crashes in house prices leave many people in **negative equity**, which leaves them unable to sell their houses (or needing to pay additional money in order to be able to do so).

Hedge funds use leverage to take huge positions in a particular asset for which they expect to see a fairly small increase in price, thus turning a small profit into a much larger one (but risking serious losses if their projections are wrong). This is why the average lifespan of a hedge fund is only about five years – many fail when highly leveraged positions go wrong, leaving their investors high and dry.

Leverage is also significant when it comes to valuing companies. Statistics such as return on equity, debt to equity and return on capital employed are all ways that investors can assess how much money a company has borrowed and how well it is investing that money. If a company is described as highly leveraged, it means that a high proportion of its investments are made with borrowed money, which can be a danger sign, as it means the company is too reliant on debt and thus too exposed to risk.

If you have outstanding debts and need to calculate the remaining balance or the appropriate loan payments, this is another example of the problem of the time value of money (see p. 117). The standard way to calculate the outstanding balance of a loan is to use the **present value of an annuity** formula. One example of this is this equation:

$$\text{Present value} = P\left[\frac{1 - (1 + r)^{-n}}{r}\right]$$

where P is the periodic payment, r is the rate per period and n is the number of periods. This is based on the assumption that (1) the periodic payment and rate don't change, and (2) the first payment is one period away.

There are a number of other variations, depending on how the loan is structured. I won't give them all here, but most basic bookkeeping guides will give the full range of equations that can be used to make these calculations.

What to Do

If you have a loan of any sort, understand how leveraged you are and what the worst case scenario is. And don't trust hedge funds too easily – some of the best-known examples are the successful ones, but there have been plenty that went out of business after their leveraged gambling led to massive losses (for the investors, not necessarily for the fund managers).

The Rule of 78

Before modern accounting systems, some notably less fair methods were used to calculate outstanding loan balances. The Rule of 78, which is now banned in the UK, the USA and many other countries, was one system which was used to weight the balance in favour of the lender.[*]

Imagine a loan of £5,000, for which the total interest over a year is agreed at £500. On a simple interest loan, the repayments will be spread evenly through the year,

---

[*] Also sometimes known as the Rule of 78s.

meaning the repayment of £5,500 is divided by 12 to find the monthly repayment. So the monthly payment is $\frac{5,500}{12}$ = £458.33 of which £41.66 is interest. If after three months you pay off the remaining balance the bank retains the first three months' interest (£125).

The Rule of 78 worked differently. In this system, for a one-year loan, the interest is stacked onto the early part of the loan by allotting 12 months' interest to the first month, 11 month's interest to the second month and so on. So the interest is divided into:

12 + 11 + 10 + 9 + 8 + 7 + 6 + 5 + 4 + 3 + 2 + 1
   = 78 segments

Similarly a two-year loan would be divided into

24 + 23 + 22 + ... + 3 + 2 + 1 = 300 segments

Now, if the borrower wants to pay back the loan after three months, the lender calculates the amount of interest that was due in that period by adding up the segments of the total interest that it deems to have fallen in those months:

12 + 11 + 10 = 33

So the borrower is charged $\frac{33}{78}$ of £500 = £211.54.

As this is getting on for half of the interest for a quarter of the period, it's easy to see why this system was grossly unfair, and was eventually discredited, although it is still unfortunately a common practice among loan sharks today. The moral of the story is never to borrow money unless you understand how the repayments are going to be calculated.

Chapter 7 Summary

1. Mathematical thought and good mental arithmetic in particular can help you to be more precise and successful in the workplace.

2. Gather the best quality data you can. And be careful not to draw conclusions from small amounts of data – beware of the effects of randomness in particular.

3. A precise visualization is an extremely useful tool. A badly constructed one can do more harm than good, so look at other people's graphs and charts carefully.

4. Game theory can provide a good understanding of the fundamentals of negotiation.

5. Devote as much of your time as possible to the most productive parts of your work.

6. Be very wary if you are borrowing money and check how the repayments will be calculated and how much you will be paying in total.

CHAPTER 8

Proving the Impossible

Alice laughed. 'There's no use trying,' she said: 'one can't believe impossible things.'
'I daresay you haven't had much practice,' said the Queen. 'When I was your age, I always did it for half-an-hour a day. Why, sometimes I've believed as many as six impossible things before breakfast.'
Lewis Carroll, *Through the Looking Glass*

Just for fun, this final chapter will look at some more complex maths puzzles and, in some cases, the prizes that are associated with them. But we should start with a health warning – some of the problems described here are fiendishly complicated. And while others are easy to understand, they are nonetheless extremely difficult to solve. It's easy to get fixated on such problems and to spend precious time struggling to come up with a brilliant new approach, but the large cash prizes mentioned are extremely unlikely to be claimed by anyone other than a professional mathematician.

Having said that, amateurs can sometimes upset the odds. For instance, consider the story of Yu Jianchun.

The Real-Life Good Will Hunting

In the Hollywood movie *Good Will Hunting*, Matt Damon plays a janitor from a rough Boston background who turns out to be a maths genius and turns his life around having been taken under the wing of brilliant maths professor Robin Williams. Depending on your point of view it's a schmaltzy but entertaining bit of fluff, or yet another movie in which Hollywood gets mathematics wrong by falling back on the cliché of the troubled genius scribbling formulae on the nearest wall or mirror.

However, there is something close to a real-life version of the story. Yu Jianchun, a Chinese migrant worker and amateur mathematician, came to international attention after inventing an alternative method of verifying Carmichael numbers. It took him eight years to persuade anyone to take him seriously, but his work has now been recognized as a brilliant achievement.

To explain his discovery, we need a very quick intro to Carmichael numbers (for those who don't already know them intimately). Pierre de Fermat's 'little theorem' states that if p is a prime number and a is not divisible by p, then

$a^{p-1} \equiv 1 \pmod{p}^*$

---

* If you need a reminder on modular maths, a number is 1 mod p if it is 1 more than any whole-number multiple of p.

This gives us a relatively quick way to test whether a randomly chosen whole number n is composite (in other words not prime). We pick values of a (that aren't divisible by n) and test for this property. If it doesn't have it, then n is definitely composite. And if, on the contrary, we find several values of a for which the statement above is true, then we have a reason to believe that n is probably prime. This is known as the **Fermat probable primality test.**

Let's work through this for a simple example:

$n = 7$ so we pick a value of a which isn't divisible by n.
$a = 6$
$p = 5$
Does $a^{p-1} \equiv 1 \pmod{p}$?
$6^4 = 1,296 = 1 \pmod 5$

OK, that is one bit of evidence, let's try another one:

$a = 10$
$p = 3$
Does $a^{p-1} \equiv 1 \pmod{p}$?
$10^2 = 100 = 1 \pmod 3$

So these two tests suggest that 7 is a prime number (which of course we already knew).

There are two problems with this approach. The first is that for some values of a and n, it is true that $a^{n-1} - 1 \pmod n$ even though n is composite. In this case *a* is called a **Fermat liar** and *n* is a **Fermat pseudo-prime**. The second problem is that, even worse, there are a tiny proportion of Fermat pseudo-primes for which **all** values of *a* that we might test are Fermat liars. These numbers are called **Carmichael numbers.** And while they represent a much smaller proportion of the number chain than primes, there is an infinite supply of them.

Personally I find Carmichael numbers easier to understand using what is known as Korselt's criterion. Apologies for the seemingly arcane references, but this one is relatively easy to grasp. A composite integer *n* is a Carmichael number if and only if its prime factors are all different to one another and, for each of its prime factors *p*, $(n - 1)$ is divisible by $(p - 1)$. For instance, the smallest Carmichael number is $561 = 3 \times 11 \times 17$. And we find that 560 is divisible by all of 2, 10 and 16.

There are only 2,163 Carmichael numbers smaller than 25,000,000,000. However, they are a real nuisance since they make the Fermat probable primality test far less useful than it would otherwise be. And while primes might seem a dry mathematician's obsession, it should be obvious from our look at digital cryptography in Chapter 6 that the ability to rapidly factorize large numbers and assess whether they are prime or composite is rather important.

This is the problem that came to fascinate Yu Jianchun. Having migrated from a rural background to the city, he worked for a delivery company but attended classes at the university when he could find the time. He had no systematic training or personal tutoring, instead thinking intensely about maths problems in his spare time and developing his own methods. It took him years to hone his idea and even longer before he could get recognition for it. The professor who finally responded to his approaches was Zhejiang University mathematics professor Cai Tianxin who described him as having 'an instinct and an extreme sensitivity to numbers'.

Pending verification, his work appears to show a completely new way of identifying whether a number is a Carmichael number or not. If so it may have extremely useful applications. Cai Tianxin is planning on publishing the theory in a forthcoming book. At this point Yu has not gained any great wealth from his discovery. However, he has become something of a local hero, and has been offered a job working in academic mathematics, which may allow him to fulfil his ambition of settling down and raising a family.

By contrast there are other outstanding problems in number theory which would be extremely profitable to anyone who made such significant progress on them.

The Beal Prize: A Real-Life Million Dollar Question

One way to make an instant million dollars would be to come up with a successful, verified proof of the **Beal Conjecture***

---

* Also known as the Tijdeman–Zagier conjecture, which some mathematicians prefer owing to the controversy over who first formulated the conjecture.

(or a counterexample to prove it is false). Unfortunately, the conjecture is a classic case of a maths problem that is reasonably easy to explain, but tremendously difficult to solve. It can be stated thus:

If
$A^x + B^y = C^z$
where A, B, C, x, y and z are positive integers with
x, y, $z > 2$
then
it must be the case that A, B and C have a prime factor in common.

For instance, the sum $27 + 216 = 243$ can be stated as $3^3 + 6^3 = 3^5$. And 3 and 6 both have 3 as a factor. Or the sum $531,441 + 4,251,528 = 4,782,969$ can be stated as $27^4 + 162^3 = 9^7$ and 9, 27 and 162 also have a common factor of 3. The conjecture states that for any solution to the equation above, A, B and C will similarly have a common prime factor.

Visualizing the Beal Conjecture

Another example of a solution to $A^x + B^y = C^z$ is 343 + 2401 = 2,744, which can be stated as $7^3 + 7^4 = 14^3$ (with 7 as a common factor).

We can represent 7^3 as a cube made up of 343 individual blocks:

$7 \times 7 \times 7 = 343$

And we can represent 7^4 as seven identical cubes:

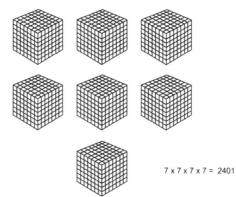

$7 \times 7 \times 7 \times 7 = 2401$

If we add these together we have eight identical cubes, which we can rearrange into a single cube, twice as big in every direction as the first cube, which thus represents 14^3:

14 x 14 x 14 = 2744

Other solutions to $A^x + B^y = C^z$ can similarly be visualized as rearrangements of cubes or sets of cubes. A counterexample to the Beal Conjecture would involve a similar rearrangement, but one in which the original cube (or cubes) had to be broken right down to the individual blocks in order to complete the final reorganization.

A version of the conjecture was formulated in 1993 by the millionaire Andy Beal, an entrepreneur and amateur mathematician who was investigating generalizations of Fermat's Last Theorem.* After Andrew Wiles proved the latter theorem in 1994, Beal decided to offer a cash prize for a proof of his own conjecture – the prize, which

* Fermat's Last Theorem is that no three positive integers a, b and c satisfy the equation $a^n + b^n = c^n$ for any integer value of n greater than 2. It can therefore be seen as a special case of Beal's Conjecture, the case in which x, y and z take the same value.

is held in a trust by the American Mathematical Society and adjudicated by the Beal Prize Committee (BPC), has been increased several times and currently stands at $1,000,000.

Beal has had a colourful career. He initially made his fortune from property investment, buying his first property for $6,500 while studying at Michigan State University. He gradually expanded his property interests, at one point buying two social housing buildings called Brick Towers in New Jersey for $25,000 and going on to sell them for over $3 million two years later. In 1988 he founded Beal Bank in Dallas, and has gone on to build it up into a business that holds over $7 billion worth of assets. He also runs Beal Aerospace, a private company that launches satellites into space.

He is a good enough poker player to have won the world's richest game of Texas Hold 'Em against a team of professional players. The minimum bet in the game, at Bellagio's casino in Las Vegas, was $100,000. And his swashbuckling approach to risk is also demonstrated by the way that Beal Bank has tended to buy assets at low points that would terrify other investors. For instance it invested heavily in power generation and infrastructure during the 2001 energy crisis in the USA, in debt instruments for the purchase of aircraft following the 9/11 attacks, and in real estate loans during the global financial crisis in 2008, when it also acquired several other banks that had failed as a result of the crisis.

However, Beal's undoubted business acumen and success have still not enabled him to find someone who can provide a solution to the maths problem which he became fascinated by over two decades ago. One reason why the problem has proved so stubborn is that it is a very broad statement, for which even special cases are difficult to attack mathematically. Fermat's Last Theorem only deals with one special case of the problem but 358 years passed from the day when Pierre de Fermat noted down the problem in the margin of his copy of *Arithmetica* before Wiles published his proof.

Peter Norvig, a director of the Google corporation, carried out an extensive search for counterexamples which showed that there are none for x, y, $z \leq 7$ and A, B, $C \leq 250,000$, and there are also none for x, y, $z \leq 100$ and A, B, $C \leq 10,000$. Of course this approach could only ever tell us whether the conjecture is true or not if a counterexample were actually found. **Failing** to find a counterexample will never prove a maths conjecture, no matter how long we go on searching, because by itself this failure will not prove that there isn't a counterexample among the infinity of cases that still haven't been checked.

It's also worth noting that we know a considerable amount about 'near misses' to counterexamples. For instance the Fermat–Catalan Conjecture concerns similar equations to those dealt with by the Beal Conjecture, where

$$a^m + b^n = c^k$$

and

$$\frac{1}{m} + \frac{1}{n} + \frac{1}{k} < 1$$

(The second condition above simply means that only one of m, n and k can take the value of 2, thus excluding the known infinitude of Pythagorean triples, such as $2^2 + 3^2 = 5^2$ or $5^2 + 12^2 = 13^2$.)

There are ten known solutions to this equation, none of which are counterexamples to the Beal Conjecture, as they all involve one of m, n or k taking the value of 2. The first few are:*

$1^m + 2^3 = 3^2$ (where m is any whole number excluding 0, 1 and 2)

$2^5 + 7^2 = 3^4$

$13^2 + 7^3 = 2^9$

and the largest known solution at the time of writing is

$43^8 + 962,22^3 = 30,042,907^2$

These solutions are naturally of interest to students of the Beal Conjecture as they provide a route by which we can try to explore what is specifically different about square

---

* $1^m + 2^3 = 3^2$ is also known to be the only solution where a, b or c is 1 and thus the only example of two consecutive integers that are powers. This was the subject of the Catalan Conjecture which was proved in 2002 by Preda Mihăilescu – it is now also referred to as Mihăilescu's Theorem as a result.

numbers that allows them to be part of such solutions while larger powers can't be.

As a moneymaking scheme it is best to regard the Beal Conjecture as a very long shot indeed. As a problem that is relatively easy to understand, the conjecture attracts a lot of attention from amateur mathematicians. Beal himself was an amateur who came to realize the magnitude of the problem. And the prize he initiated has been a magnet for cranks and eccentrics who think they might just be the one to find a solution that has evaded so many more experienced mathematicians.

A final warning about the Beal Conjecture: like many apparently 'simple' number theory problems, this is a conjecture that can be seriously addictive for those who set out to understand it. Simon Singh's book *Fermat's Last Theorem* details ways in which that problem has entranced the imaginations of amateur and professional mathematicians for centuries, inspiring the kind of passion that can lead to intrigue, deception or even madness.

As a broader, more difficult version of the same problem, and one with a million dollar price tag to boot, the Beal Conjecture is likely to lead to many more such cases of obsession in the future.

What Not to Do

Don't get so obsessed with trying to prove the Beal Conjecture that you forget to pursue other moneymaking opportunities. It took over 350 years to prove Fermat's Last Theorem, and it could easily take another 350 years before we know if the Beal Conjecture is true or not.

The Millennium Prize Problems – A Brief Introduction

The Clay Institute was created in 1998 by the philanthropist Landon Clay, a successful businessman with a background in finance and venture capital funding. He wasn't a mathematician but had a keen interest in the subject and a conviction that the importance of mathematics in society was underappreciated. The goal of the institute is to support and promote mathematical research and to encourage new breakthroughs in mathematical knowledge.

Towards this end, the institute listed seven important outstanding mathematical problems in 2000. A verified solution to any of these problems would win $1,000,000 (a Millennium Prize) from the institute. The only Millennium problem that has been solved to date is the Poincaré Conjecture. Of course these are all problems that are unlikely to be solved by any but the most expert of mathematicians, so it's probably best not to get your hopes up too high. But, for the record, here's a very brief summary of the problems.

THE RIEMANN HYPOTHESIS

This hypothesis is named after Bernard Riemann, who proposed it in 1859. It is regarded as one of the most important unsolved problems in mathematics. To understand roughly why, you need a grasp of the prime number theorem, which describes the way the density of prime numbers decreases as you move up through the whole numbers. It tells us that for large numbers (n), the probability that a random integer is prime is very close to $\frac{1}{\log(n)}$. However there is a stubborn error term in the theorem which means that it is always slightly inaccurate.

The Riemann hypothesis concerns a particular function (the 'Riemann zeta function') which can be applied to complex numbers (which have a real and imaginary part) to find both 'trivial' and 'non-trivial' zeros. The Riemann hypothesis is that that all the 'non-trivial' zeros of the zeta function are complex numbers with real part $\frac{1}{2}$.

That sounds pretty confusing and it is, but I'll try to unpack it slightly. A function takes a value and outputs a result. For instance if you input 3 into the function $f(x) = x - 3$ then the output is zero. This is the only zero of that function, but some functions have many zeros.

This is the real number version of the zeta function

$$\sum_{n=1}^{\infty} \frac{1}{n^s} = \frac{1}{1^s} + \frac{1}{2^s} + \frac{1}{3^s} + \frac{1}{4^s} + \dots$$

A real number is any rational or irrational number, for instance 1, –7, $\frac{1}{5}$, pi or the square root of 2. But mathematicians like to make things more complicated so there is a way of converting this into a slightly different function which allows complex numbers to be input. Complex numbers are a combination of real and imaginary numbers: for example $3 + i$ (where i is the square root of –1, something which is impossible in our ordinary number system) is a complex number. When we put complex numbers into the modified zeta function, there are some zeros which are obvious and easy to find (the 'trivial' zeros) which are of little interest, but also some zeros that are harder to find and which are extremely interesting (the 'non-trivial' zeros).

Riemann used this idea to develop a more precise way of estimating the number of primes below a certain number n – the importance of the Riemann zeta function is that the magnitude of the oscillations of primes around their expected position is related to the real parts of the zeros of zeta function. Again, this is complicated to explain, but the basic point is that Riemann used the function to develop a new, much more accurate way of expressing a formula to predict the number of primes up to a given number, meaning that the function in some way 'encodes' information about the position where we would expect the prime numbers to be found. Most mathematicians believe that the Riemann hypothesis is correct, and there are many potential 'proofs' of other results which would hold if and only if the hypothesis is correct – so a proof of the hypothesis would immediately validate a lot of other mathematical theory. If you want to read more about it, *The Music of the Primes* by Marcus du Sautoy is the best non-technical explanation I have come across.

THE P VS NP PROBLEM

There is a difference between finding a solution to a difficult mathematical problem, and checking whether a given solution is correct. This difference lies at the heart of the P vs NP question, a crucial unsolved problem in computer science. The question is whether all problems that can be 'quickly' verified can also be 'quickly' solved. 'Quickly' here has a technical meaning which is 'in polynomial time' – in other words does the amount of time required to solve the problem vary in proportion to the amount of input data as a polynomial function, or, for instance, as an exponential function? Think for instance about the speed that these progressions grow in size.

10^n: 10^1, 10^2, 10^3, ... = 10, 100, 1,000, ...

n^2: 10, 100, 10,000, 1,000,000, ...

m^n: m^{10}, m^{100}, m^{1000}, ...

Hopefully it is fairly obvious that a problem whose size

was prone to expanding in proportion to the terms in the third series, where the exponent is growing in proportion to the first series would quickly become impossibly complex. Out of the problems which can be verified 'quickly', those for which the solution can be found 'quickly' are known as 'P' while those for which the solution can't are known as (NP). An example of a problem that seems to be an NP problem is the Hamiltonian Path Problem: if you have N cities to visit, is it possible to do this without visiting a city twice (see Figure 43)? This is a very hard problem to solve (for large networks) but once a solution is found it is fairly simple to check that it works.

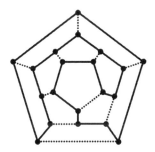

Figure 43. The challenge is to visit each node in the network once only – the bold lines indicate the path taken, which is easy to verify but (relatively) hard to compute.

A proof either way of the P vs NP problem would have significant implications for cryptography, mathematics, artificial intelligence, and many other fields of study.

BIRCH AND SWINNERTON-DYER CONJECTURE

This conjecture is a description of the set of rational solutions to equations defining an elliptic curve. It is named after mathematicians Bryan Birch and Peter Swinnerton-Dyer, who developed the conjecture in the 1960s. While a few special cases have been proved, the wider conjecture remains unproven, although there is a lot of evidence that it is probably correct. We saw on p. 216 that elliptic curves are crucial with respect to digital cryptography, and they can be used in the factorization of numbers into prime factors, so any advance in our understanding of them may be significant when it comes to the future of online security.

YANG–MILLS AND MASS GAP, THE NAVIER–STOKES EQUATION, THE HODGE CONJECTURE

The first three problems I have described are the easiest of the Millennium prize problems for someone without advanced mathematical knowledge to at least dimly understand. The next three are pretty intense and I won't pretend to have a perfect grasp of any of them, but here is a quick overview.

The Yang–Mills equations are part of our basic understanding of particle physics. I'm told that experimental simulations suggest there is a 'mass gap' in the solution to quantum versions of the equations but there is no proof of this property. So this is known as 'Yang–Mills and the mass gap'. If that sounds like gobbledygook to you, don't worry about it – it sounds like gobbledygook to me too and I won't pretend to be able to explain it.

The Navier–Stokes Equation governs the flow of fluids including air and water through space. There are many practical applications for this equation. However, there are significant gaps in our understanding of the equation, especially when it comes to turbulence, which modern physicists don't have a satisfactory way of modelling. So a solution to this problem would be a significant advance in physics and our understanding of the physical world.

Topology is the study of shapes and how they can be transformed into one another. The Hodge Conjecture is an unsolved topological problem, regarding how much of the solution set of a system of algebraic equations can be defined in terms of further algebraic equations. It has been proved for cases in less than four dimensions, but not for four dimensions.

(I can give the reader a cast-iron guarantee I won't be winning a million dollars by solving any of these three problems...)

POINCARÉ CONJECTURE

Finally, this conjecture, formulated in 1904 by the French mathematician Henri Poincaré, is the only one of the Millennium prize problems that has actually been solved. The conjecture, which was one of the longest-standing problems in topology, was that a three-dimensional sphere is identical to a three-dimensional manifold (a defined set of points) given certain algebraic conditions. It was proved by the Russian mathematician Grigori Perelman in 2006. Perelman was offered the Fields Medal and was also eligible to receive $1 million from the Clay Institute – however, he declined both the award and the prize on the basis that other mathematicians had paved the way for his proof and it would be unfair for him to accept it.

His refusal to take such prizes is a reminder that most mathematicians are more interested in academic achievement than material wealth, and that most of the ways in which you are likely to enhance your wealth rely on the everyday fundamentals of maths rather than such complex abstractions.

Other Maths Prizes

Maths prizes are, realistically, aimed at professional mathematicians, though we can all daydream. The most prestigious prize for professional mathematicians is the Fields Medal. The prize is a mere 15,000 Canadian dollars, but this prize really marks out a mathematician among their peers. It is also restricted to people under the age of 40 – the mathematician John Charles Fields who established the prize in 1936 wanted it to be an encouragement to the best younger mathematicians and a stimulus to their ongoing careers.

There are, however, some prestigious prizes that are more lucrative. The Chern Medal is named after the late Chinese mathematician Shiing-Shen Chern and is awarded by the International Mathematical Union every four years, with a cash prize of $250,000. Even more financially rewarding is the Abel Prize – this is awarded annually by the government of Norway and is worth 6 million kroner (about £575,000). It's named after the nineteenth-century Norwegian mathematician Niels Henrik Abel – there was an initial attempt to establish the award in 1901, when it became apparent that the newly created Nobel prizes didn't include a category for mathematics. The initial momentum fell through, but interest in the idea was rekindled in 2001 and it has now become a significant prize in the field.

The reason why there is no Nobel prize for mathematics is not entirely clear. A story has been circulated that the leading Swedish mathematician Gosta Magnus Mittag-Leffler had run off with Alfred Nobel's wife, and Nobel refused to create a maths prize in revenge. However, given that Nobel never actually married, an alternative version that has also been published is marginally more likely to be true – this is that the wealthy Mittag-Leffler had crossed Nobel in business. In the end, though, the truth may be much more boring. There is no proof that the two men had anything much to do with each other and it may well be that Nobel simply wasn't that interested in mathematics and it didn't even occur to him to create such a prize.

Now that the Abel Prize has filled the gap, it has created a path to considerable wealth for a few lucky and talented mathematicians.

The Collatz Conjecture and Other 'Easy' Unsolved Problems

We have seen that formal prizes are a distant prospect for all but a few academics. It is, however, worth mentioning a few more 'easy to state, hard to solve' problems which often fascinate amateur mathematicians. We've seen how Yu Jianchun contributed significantly to one outstanding problem, so it is not impossible that someone may one day solve one of the problems below and be rewarded with acclaim and (perhaps) monetary reward. However, be warned, these problems are all seductive if you have a problem-solving mind – it is relatively easy to play with them and to start to come up with ideas as to how they might be proved, but all have remained stubbornly unsolved over the years, and it may be that they are simply impossible to prove.

THE COLLATZ CONJECTURE

Take any positive whole number. If it is even, halve it, if it is odd, multiply by 3 and add 1. You will end up with a chain of numbers like this:

7–22–11–34–17–52–26–13–40–20–10–5–16–8–4–2–1

Thereafter you will continue to cycle around $1 \rightarrow 4 \rightarrow 2 \rightarrow 1$. The Collatz Conjecture (or $3n + 1$ Conjecture) suggests that for any starting number you will end up reaching 1. There are no known counterexamples but, of course, this doesn't mean that one might not be found eventually.

The problem is an interesting one because there are a few counterexamples to slightly different versions. If you allow negative numbers as starting points (with –1 as the usual end of the chain) then we have these cycles, which will never reach –1:

$$(-7) \to (-20) \to (-10) \to (-5) \to (-14) \to (-7)$$
$$(-17) \to (-50) \to (-25) \to (-74) \to (-37) \to (-110) \to (-55) \to$$
$$(-164) \to (-82) \to (-41) \to (-122) \to (-61) \to (-182) \to$$
$$(-91) \to (-272) \to (-136) \to (-68) \to (-34) \to (-17)$$

Similarly if we try the conjecture using $5n + 1$ for odd numbers instead of $3n + 1$ we find counterexamples starting at 13 and 33. It is also possible that, instead of a number that cycles back to itself without reaching 1, there might be a starting number that leads to an infinitely rising series.

It's a fun conjecture to play around with, but notoriously difficult to make any progress with. It's one of numerous problems for which the brilliant mathematician Paul Erdős promised prizes (and which his estate might still pay out on). However, he only offered $500 and his opinion was that it was probably impossible, saying 'Mathematics may not be ready for such problems.' Possibly the greatest living expert on the problem is Jeffrey Lagarias who suggested in 2010 (after decades of study) that it may be 'unprovable'.

GOLDBACH'S CONJECTURE

This is another conjecture that is extremely easy to understand – it is simply that every even number greater than 2 can be expressed as the sum of two primes. The German mathematician Christian Goldbach suggested this idea in a letter to Leonhard Euler in 1742 and it has remained stubbornly unproven ever since.

It seems extremely likely that it is true, since the number of 'Goldbach partitions' seems to increase as you move up the number chain. (A Goldbach partition for the number n is a way of adding two primes together to make n. So the number 24 has the Goldbach partitions $5 + 19$, $7 + 17$ and $11 + 13$.)

There has been some progress towards a proof, but no decisive final step. For instance it has been shown that every number greater than 4 is the sum of at most four primes.

Beware if you come across a story about a million dollar prize for a proof of this conjecture – this was offered in 2000 by the UK publisher Faber & Faber, but it was part of a publicity campaign for the book *Uncle Petros and Goldbach's Conjecture* and has now been withdrawn. A genuine proof would certainly bring fame to its discoverer, but after more than 270 years, that proof may continue to remain elusive.

TWIN PRIMES

Prime number conjectures are a bit of a magnet for geeks like me. The Twin Prime Conjecture is actually quite closely related to Goldbach's Conjecture as it also concerns pairs of primes and the patterns of these pairs.

The proof that prime numbers are infinite goes all the way back to the Greeks – Euclid's wonderful proof simply asks us to imagine that there is a largest prime number, then to multiply all the prime numbers together and add 1. Since this must be either a new prime or a product of primes that weren't in our original list, it is impossible that our original assumption was correct. So there is no largest prime number and they go on forever.

In recent centuries mathematicians have become fascinated by a related problem – the question of whether twin primes (where p and $p + 2$ are both prime) are also infinite. It seems like we should be able to use a similar method to Euclid, but in practice this simply doesn't work. However, there has been some progress. In 2013 the Chinese mathematician Yitang 'Tom' Zhang revealed a proof that states there are infinitely many pairs of prime numbers that differ by 70 million or less. This may seem a bit weak but it was nonetheless a huge step forwards, since it showed there are gaps for which an infinite number of pairs exist. The gap has been reduced from 70 million down to 246 in subsequent work, and may yet reach 2. But, as things stand, this (and the related set of conjectures about cousin primes (p and $p + 4$), sexy primes (p and $p + 6$) and so on) remains a fascinating unsolved conundrum.

There are plenty more unsolved problems out there: the list of conjectures relating to prime numbers alone is a long one. Though if you do want to delve into such areas, then it is probably best to stick to doing it for fun rather than for financial reward.

Codebreaking

The art of cryptanalysis (codebreaking) has always appealed to the mathematically minded – many of the great codebreakers of history came from a maths background. A solution to any of the most famous unbroken codes from the past would potentially bring fame and even fortune to the codebreaker.

As an example of what can be achieved, the famous Copiale cipher was a mystery that lasted more than 270 years. The cipher was contained in a manuscript of 75,000 handwritten characters. The code was finally cracked in 2011 by a team consisting of Kevin Knight of the University of Southern California and Beáta Megyesi and Christiane Schaefer of Uppsala University in Sweden. The mathematical methods they used, including frequency analysis, showed that the document used a complex homophonic cipher (one in which different ciphertext letters can be substituted for the same plaintext letter) and was the annals of a secret society of Oculists (a kind of Masonic group of ophthalmologists) from the 1730s.

So persistence, luck and good mathematical methods can in theory be used to crack such ancient mysteries. One example of a text that remains mysterious is the Voynich manuscript. This is a fascinating and beautifully decorated book containing diagrams that appear to be botanical or astrological, and a text in an unknown language. It is 240 pages long, and contains 170,000 characters of which about 30 occur the most regularly. Not much is known about the book; its age is uncertain and many of the botanical drawings remain unidentified. Decades of study have failed to provide a way of deciphering the manuscript, although some of its students do claim to have deciphered short sections. It has been suggested that rather than being a traditional cipher it may be a piece of steganography in which a grid placed over the page would reveal which letters to decode, which would only make it a more difficult problem to solve.

Another unbroken cipher is the D'Agapeyeff cipher (Figure 44) which appears in the first edition of *Codes and Ciphers*, a cryptography manual written by Alexander D'Agapeyeff in 1939. It was presented without a solution as a challenge to the reader – its brevity means that it may never be cracked. It was omitted from later editions of the book and D'Agapeyeff admitted to having forgotten how he had encrypted it in the first place.

```
75628 28591 62916 48164 91748 58464 74748 28483 81638 18174
74826 26475 83828 49175 74658 37575 75936 36565 81638 17585
75756 46282 92857 46382 75748 38165 81848 56485 64858 56382
72628 36281 81728 16463 75828 16483 63828 58163 63630 47481
91918 46385 84656 48565 62946 26285 91859 17491 72756 46575
71658 36264 74818 28462 82649 18193 65626 48484 91838 57491
81657 27483 83858 28364 62726 26562 83759 27263 82827 27283
82858 47582 81837 28462 82837 58164 75748 58162 92000
```

Figure 44. The D'Agapeyeff cipher.

In theory, one of the most potentially rewarding unbroken bits of encryption is the Beale ciphers. These come from a pamphlet published in Virginia in the 1880s, containing a story and three encrypted messages. The story claims that, decades earlier, a man named Beale buried two wagons of treasure at a secret location in Bedford County, Virginia. He is supposed to have left a locked box at a local inn, then left town for good. Many years later the innkeeper opened the box to discover the messages. After the innkeeper died, a friend of his took twenty years to decrypt just one of the messages, which described the gold, silver and jewels that were buried. The remaining messages apparently give exact directions to the location of the treasure, which could still be recovered by anyone who solves the ciphers.

A word of warning though. There is no reliable record of the man called Beale ever having existed and it may well be that the whole story was a piece of publishing schtick dreamed up to sell pamphlets. So while solving the ciphers may be the path to riches, it's perhaps more likely that the best moral to take from the story is that nothing sells like a treasure hunt, even one that might turn out to be a wild goose chase.

Famous Problems that were Solved

This last chapter has discussed a variety of maths problems and codes that may never be cracked. But let's not end the chapter on too negative a note. As we've mentioned, stubborn maths problems do sometimes get solved.

We've seen how a team of academics cracked the Copiale cipher after 270 years, how Grigori Perelman proved the Poincaré Conjecture a century after it was formulated, and how Andrew Wiles has gained worldwide fame for his (extremely complex) proof of Fermat's Last Theorem after 358 years. And there have been plenty of other cases where longstanding problems have finally been solved.

For instance in 1919 the Hungarian mathematician George Pólya conjectured that 50% or more of natural numbers (also known as the counting numbers, 1, 2, 3, ...) less than a given number would have an odd number of prime factors (rather than an even number). The so-called Pólya Conjecture was proved false in 1958 by C. Brian Haselgrove. It's an interesting example of why not yet having found a counterexample cannot be considered a proof of a conjecture – when Haselgrove first disproved this one, his work only showed that there must be a counterexample somewhere in the region of 1.845×10^{361}! Eventually a much smaller counterexample ($n = 906,180,359$) was discovered.

Another settled problem that is relatively easy to understand is the four-colour theorem. Proposed in 1852 by F. Guthrie, this was the suggestion that any map in a plane (a two-dimensional surface) separated into different regions can be coloured using only four colours in such a way that no two adjacent regions are the same colour. (If regions share a stretch of boundary longer than a single point, they are regarded as adjacent.)

The first non-flawed proof of this problem came in 1977 when two mathematicians put forward a computer-assisted proof that it was always feasible to colour the map with only four colours. (Some of their peers rejected this proof on the basis that it involved computer analysis of a list of examples, but it has since been confirmed by a separate independent proof that doesn't rely on such means.)

So, longstanding problems can be solved. Prizes can be won. Treasures can be found. It may not happen every day, but it never does any harm to dream of being the one to finally find your way through the labyrinth to uncover the solution that lies at its heart.

Chapter 8 Summary

1. Outstanding problems in mathematics or cryptanalysis can sometimes be solved, even after centuries.
2. Yu Jianchun's work on Carmichael numbers has shown that even today an amateur mathematician can make a major breakthrough.
3. More realistically, pondering complex conjectures is a good way of sharpening your mathematical instincts and (for some) an enjoyable pastime.
4. There may be some buried treasure out there, but most of us are going to have to find better ways to make a million.

CONCLUSION

Being Maths Aware

In this book, we've looked at a few different aspects of the links between mathematics and wealth. We've explored the world of gambling, and seen how the insights from that sphere can be applied more profitably to speculation and investment. We've jogged through the basics of how to use portfolio theory to spread risk in a portfolio of investments. We've taken a look at the mathematical basis for the vagaries of the stock markets. We've recounted ways that some people have hacked or gamed the system using their mathematical insights. We've seen how quants and algorithms have taken over the financial world, and had a peek at the future of mathematics, cryptography and business. We've considered the best ways to carry over mathematical insights into the workplace to improve your performance there. And we've spent a while daydreaming about maths prizes, unsolved ciphers and centuries-old conjectures.

My conclusion is that there genuinely are some ways to make a million using mathematics. Some of these are more achievable than others, most require hard work and long hours of practice as well as maths ability, and some routes are only available to professional mathematicians. But anyone can dream and work to try to use their mathematical skills in the most profitable or enjoyable way possible.

Self-help books, especially those that focus on getting rich quick as a path to happiness, are fond of advising the reader to write down their goals. In the case of mathematical thinking, I think it is worth making a list of the things in this book that could genuinely make a difference in your life. Perhaps you don't have the inclination to go after big money prizes and the patience to learn portfolio theory? If so, it still might make a big difference to you to become more aware of statistical distortions and to study game theory before it comes to your next salary negotiation. Perhaps you can't see any loopholes in your local lottery or the latest game show? But you can still understand how randomness affects your daily life, and avoid irrational fallacies in your investment decisions.

Have a think about what you can genuinely achieve and then think positively about your ability to make a difference. Self-made millionaires didn't all become rich from a single good idea. Many of them did well by focusing on a range of different aspects of their approach to business and life and by trying to minimize every loss and maximize every profit.

Being maths aware in your approach to business, investment and speculation can make you significantly more likely to avoid mistakes, to retain and invest your money more effectively and to be more effective on a day-to-day basis. Numbers and money have been inextricably linked since the first economic transactions in early history, and it is as true today as it ever was that a healthy understanding of mathematics is a crucial foundation stone for financial success.

Hugh Barker is a non-fiction author and editor; as the latter he has edited several successful popular maths books, including *A Slice of Pi* by Liz Strachan. Hugh is a keen amateur mathematician, and was accepted to study mathematics aged 16 at the University of Cambridge.